PEOPLE PERSON

WOMEN'S ADVENTURES IN SCIENCE

PEOPLE PERSON
the story of sociologist
MARTA TIENDA

by Diane O'Connell

Franklin Watts
A Division of Scholastic Inc.
New York • Toronto • London • Auckland • Sydney
Mexico City • New Delhi • Hong Kong
Danbury, Connecticut

Joseph Henry Press
Washington, D.C.

AUTHOR'S ACKNOWLEDGMENTS

My greatest thanks to Marta Tienda for educating me about her work, for opening up her life to me, and for the hours she took out of her busy schedule to patiently answer my many questions. Marta's family also generously gave of their time for extensive interviews: Toribio Tienda, Maggie Chavez, Irene and Rubén Rumbaut, and Gloria Trevino. Many other friends and colleagues graciously made time to answer questions, especially Drew Altman, Frank Bean, Jeanette Bowman, Harley and Waldi Browning, Vartan Gregorian, Grace Kao, Doug Massey, Jeff Morenoff, Lucille Page, Emilio Parrado, Dudley Poston, Mike Rosenfeld, Rudee Rossi, Haya Stier, Teresa Sullivan, James Trussell, and William Julius Wilson. A special thanks to Diana Sacke for her responsiveness to my myriad requests. I also thank Billy Goodman for his extensive research at the beginning of this project. Finally, I'd like to thank my husband, Larry, for his moral and creative support and his unflagging enthusiasm for my work. — DO'C

Cover photo: Sociologist Marta Tienda is shown in her office at the University of Chicago.

Cover design: Michele de la Menardiere

Library of Congress Cataloging-in-Publication Data

O'Connell, Diane, 1956-
   People person : the story of sociologist Marta Tienda / Diane O'Connell.
      p. cm. — (Women's adventures in science)
   Includes bibliographical references and index.
   ISBN 0-531-16781-X (lib. bdg.) ISBN 0-309-09557-3 (trade pbk.) ISBN 0-531-16956-1 (classroom pbk.)
   1. Tienda, Marta. 2. Sociologists—United States—Biography—Juvenile literature. I. Title. II. Series.
   HM479.T54O36 2005
   301'.092—dc22

                                                                        2005000825

Any opinions, findings, conclusions, or recommendations expressed in this volume are those of the author and do not necessarily reflect the views of the National Academy of Sciences or its affiliated institutions.

Printed in Mexico.
   2 3 4 5 6 7 8 9 10 R 14 13 12 11 10 09 08 07 06

About the Series

The stories in the *Women's Adventures in Science* series are about real women and the scientific careers they pursue so passionately. Some of these women knew at a very young age that they wanted to become scientists. Others realized it much later. Some of the scientists described in this series had to overcome major personal or societal obstacles on the way to establishing their careers. Others followed a simpler and more congenial path. Despite their very different backgrounds and life stories, these remarkable women all share one important belief: the work they do is important and it can make the world a better place.

Unlike many other biography series, *Women's Adventures in Science* chronicles the lives of contemporary, working scientists. Each of the women profiled in the series participated in her book's creation by sharing important details about her life, providing personal photographs to help illustrate the story, making family, friends, and colleagues available for interviews, and explaining her scientific specialty in ways that will inform and engage young readers.

This series would not have been possible without the generous assistance of Sara Lee Schupf and the National Academy of Sciences, an individual and an organization united in the belief that the pursuit of science is crucial to our understanding of how the world works and in the recognition that women must play a central role in all areas of science. They hope that *Women's Adventures in Science* will entertain and enlighten readers with stories of intellectually curious girls who became determined and innovative scientists dedicated to the quest for new knowledge. They also hope the stories will inspire young people with talent and energy to consider similar pursuits. The challenges of a scientific career are great but the rewards can be even greater.

Other Books in the Series

Beyond Jupiter: The Story of Planetary Astronomer Heidi Hammel

Bone Detective: The Story of Forensic Anthropologist Diane France

Forecast Earth: The Story of Climate Scientist Inez Fung

Gene Hunter: The Story of Neuropsychologist Nancy Wexler

Gorilla Mountain: The Story of Wildlife Biologist Amy Vedder

Nature's Machines: The Story of Biomechanist Mimi Koehl

Robo World: The Story of Robot Designer Cynthia Breazeal

Space Rocks: The Story of Planetary Geologist Adriana Ocampo

Strong Force: The Story of Physicist Shirley Ann Jackson

Contents

Introduction   ix

1. Sociology of the Soul   1
2. Toby's Dream   7
3. On the Move   11
4. Shelter from the Storm   19
5. A Dawning Dream   25
6. Marta's Mission   31
7. Bonus Days   37
8. Enter the Maestro   45
9. Family Matters   59
10. Asking Why   69
11. The Immigrant Experience   77
12. Marta's Dream   85

Timeline   94
About the Author   96
Glossary and Metric Conversion Chart   97
Further Resources and Bibliography   99
Index   101
Series Advisory Boards   105
Credits   107

Life's Work

Marta Tienda loves numbers. She also loves people. How are these two related?

Numbers, Marta knows, can answer questions about people—questions such as: Why are some people rich and others poor? Why do some people have good jobs while others have low-paying ones? Why do some people have a better chance of getting a college education than others?

Marta asks these questions because she's curious, but she also wants to find solutions. Marta is a sociologist, a scientist who studies the behavior of groups, organizations, and societies. You might even consider her a kind of detective.

Just as a detective pieces together clues to re-create a picture of a crime, Marta searches for clues in numbers to form a picture of a group of people. And just as a detective's work helps solve personal crimes, Marta's work helps solve social injustices: why certain people get more in life than others.

Marta Tienda knows the pain of having less. She grew up poor, the daughter of a Mexican immigrant who never had a chance to finish elementary school. But she beat the odds and became one of the country's top sociologists. And ever since she was a college student, Marta has used social science to improve the lives of other people—people just like her.

In their faces

Marta saw reflections

of her own *father*.

SOCIOLOGY OF THE SOUL

It's *going to be a hectic day*, thought Marta Tienda. It was July 1971, and Marta was standing behind the counter where she worked in the basement of the Alpena County government building in Alpena, Michigan. She gazed out over the long line of men, women, and children snaking through the office as they waited to be certified (approved) for food stamps. Almost all of them were Mexican American migrant workers from Texas.

Many of the workers had been traveling with their families from South Texas to this northern region of Michigan for years, often picking crops on the same farms summer after summer. In their faces Marta saw reflections of her own father as a younger man, standing in line just for the chance to do a hard day's work at low wages. Speaking little English—and lacking both job skills and schooling—her father had crossed into the U.S. illegally from his home in Mexico in 1941.

Marta's job was to decide whether each person deserved United States government food stamps, which could be used to buy food at supermarkets. Workers who earned less money or had bigger families were eligible to receive more food stamps.

Marta had just finished her junior year of studies at Michigan State University. She was excited about this summer job.

Like these Mexican farm laborers crossing the Rio Grande River *(above)*, Marta's father, Toribio *(opposite)*, came to the United States illegally to find work picking crops.

A woman uses food stamps to buy groceries. Each "food coupon" has a certain dollar value that can be exchanged for food.

Working at the state agricultural department gave her a chance to interact with people in need.

Next up at her counter was a young woman who couldn't have been much older than Marta, who would turn 21 the next month. A baby nestled in the young woman's arms, and a five-year-old girl clung to her side. The little girl's curly black hair and big brown eyes reminded Marta of herself as a young girl, when she and her family had picked tomatoes, cucumbers, and cherries to make ends meet.

Speaking in Spanish, Marta asked the woman questions that would tell her the amount of food stamps to approve.

"Did you work in the last month?"

"No," the young mother replied in a soft, hesitant voice. If she had answered "yes," Marta would have then asked the woman how much she earned.

"How many people in your family?"

"Six." It was common for migrant workers to have large families.

Marta then told the woman that she was eligible for food stamps. Because the woman had not earned any income in the previous month, the stamps would be free.

Marta loved her work because she could make decisions that improved the way people lived. Though she didn't yet know it, that summer put her on the path to becoming a social scientist—a career that would give her the opportunity to make a difference in the lives of many more people.

~ Expanding the American Dream

The United States was built on a democratic ideal—the ideal that all people are created equal and that they have the right to life, liberty, and the pursuit of happiness. In other words, anyone who works hard and plays by the rules deserves to lead a good life.

In reality, that's not always the case.

Why is that?

Many social scientists gather facts to show that not everybody enjoys the same chance of achieving the American dream. A social scientist might say, "Let's figure out how many people are in this situation. How many are cut off from having an opportunity to live the American dream? Is it a third of the population? Is it more? Less?"

> **The United States was built on a democratic ideal—the ideal that all people are created equal and that they have the right to life, liberty, and the pursuit of happiness.**

Finding out why certain groups are disadvantaged is an important step—that way, the people who govern can decide how to remedy it.

Suppose we find that Anglo-Americans (those with mainly European backgrounds) make more money than Mexican Americans.

The American Dream has existed throughout the history of the United States, and it continues to guide people today. Marta's father strongly believed in an important part of the dream: That with hard work and discipline, his children could grow up to live better lives than he did.

Mexican immigrants who came to the United States illegally often worked long hours picking crops. These laborers are picking cantaloupes in California.

Is that because people of Mexican descent are deliberately kept out of jobs that pay well? Or is it because many of them are recent immigrants, or the children of recent immigrants, and are still getting their education?

Marta Tienda's career has centered on how to make the American dream a reality for more racial and ethnic populations, particularly Hispanics. She focuses primarily on how to provide greater opportunities in education and employment.

~ The Power of the Personal

Marta's humble origins have motivated her entire career. As a social scientist, she brings the power of personal knowledge to her work. In fact, many of the questions she asks rise directly from her own experience.

When she probes the history of Mexican Americans, for example, she is exploring her own past.

When she looks at the effects of poverty on families, she is remembering her own childhood of need.

When she considers what it's like for a single parent to raise a family, she is reflecting on her mother's death when Marta was just six years old, leaving her father to raise five children by himself.

Most of all, when Marta Tienda seeks to find out why some people make it and others do not, she recognizes the power of a parent's drive for his or her child to succeed. Without her father's determination, Marta knows, she would not be as successful as she is today. Intent that his children would finish high school, her father set the standard for excellence. He encouraged them to do their schoolwork and never missed a parent-teacher meeting for Marta or her brother and three sisters.

Marta's father had a dream of a better life for his family—a dream that began with a perilous journey.

Toribio Tienda, with Marta *(left)* and her older sister Maggie, was determined that his children would receive the education he could not get for himself.

Toby and Azucena
wanted to *give*

their children a **better life.**

TOBY'S DREAM

2

Toribio Tienda huddled against the chill night air, crowded onto a makeshift raft with his uncle and a "coyote"— a person paid to help immigrants cross borders illegally. It was 1941, and many Mexicans were hoping to find work at U.S. farms and factories where workers had gone off to fight in World War II.

With only one paddle to steer the canoe-like raft, they crossed the Rio Grande from Mexico into Texas. It was a dangerous trip. Some Mexicans drowned while crossing; others were cruelly shot by snipers firing from the American shore. Still others got turned back or jailed by the U.S. Border Patrol.

But for the 16-year-old Toribio Tienda, known as Toby, the risks were worth taking. He and one of his uncles hoped to find work on a farm or ranch in South Texas to earn money for their hungry families back home.

Toby had left his parents and his nine brothers and sisters in Hualahuises [hwa-la-HWEE-sus], a dry, dusty town in northeastern Mexico. There, on their tiny farm, his family grew corn and beans and raised goats and chickens. They had no indoor plumbing or electricity. Because the children had to work, they couldn't go to school; Toby had made it through only the fourth grade.

Toby *(above)* moved his family from Texas to Michigan to seek a better life. During a stroll in a park in Detroit *(opposite)*, Toby holds Marta's hand as Maggie keeps pace alongside them.

Like these migrant workers loading baskets of spinach onto a truck, Toby worked hard picking grapefruits and oranges when he came to Texas.

Once safely in Texas, Toby found work picking grapefruits and oranges. He also drove a tractor and planted fruit trees. Toby's boss gave him a place to live and let him eat whatever the farm produced, but he had to buy other food out of his wages of $1 a day.

~ Secret Marriage

Lacking legal immigration documents, Toby ran the risk of being picked up and deported (sent out of the country). Yet so many jobs needed to be filled in the United States that immigration officials often looked the other way when illegal aliens were employed.

Next, Toby moved to the Rio Grande valley town of Edcouch, Texas. There he found work in a fruit-packing house that paid big money for illegal aliens at that time: 45 cents an hour.

One day he and a Mexican girl named Azucena started tossing grapefruits at each other. As the weeks went by, their flirting blossomed into a romance, and in the fall of 1946 the two got married in secret. They had to do it that way because both Toby and Azucena knew that her parents would never approve of him; they had wanted their daughter to marry a man with more money.

The newlyweds settled into Toby's small rented house in Edcouch. Because Azucena was a U.S. citizen—she had been born to Mexican migrant workers in Michigan in 1929—Toby now became entitled to reside in the United States. As soon as the fruit-picking season ended, he used his new legal status to find work on the Missouri Pacific Railroad, where

Azucena poses with Maggie and newborn Marta.

he earned a much higher salary than he could packing grapefruits. Soon he and Azucena had a daughter, Maggie. Two years later, on August 10, 1950, a second daughter was born. Her name: Marta.

~ The Quest for a Better Life

Toby and Azucena wanted to give their children a better life. That meant getting better jobs. But how? And where? Toby's Uncle Joe worked in a steel mill in Detroit. Maybe he could help Toby get a job there too? Though he had never met his Uncle Joe, Toby and Azucena decided to move north; they wanted better jobs for themselves and better lives for their two young daughters, Maggie and Marta. It was common then, as it is now, for immigrants to get help from family members who were already established—even if they didn't know one another very well.

In the early spring of 1951, the young family of four boarded a bus with all their possessions in tow. Detroit, here we come!

The Tiendas' excitement evaporated when they woke up to the reality of living conditions in Detroit, which were not much better than they had been in rural Texas. After living briefly with Uncle Joe, they moved into the only place of their own they could afford—a cramped basement apartment in a run-down neighborhood.

Marta takes a walk in her new neighborhood, a run-down part of Detroit where the family lived for about a year and a half.

Toby began working at Great Lakes Steel, which made parts for the auto industry. After a long day spent working on high, hot electrical cranes, Toby would gratefully unhook his heavy tool pouch and trudge home, bone-tired.

A year after Marta's birth, a younger brother, Juan Luis, arrived. The Tiendas' tiny two-bedroom apartment was now packed to the bursting point. Something had to give.

Marta was still
learning English,

and she felt **bewildered**
most of the time.

ON THE MOVE

In August of 1952, when Marta turned two, the Tiendas moved to a bigger apartment in a better neighborhood. They were now in public housing—apartments for low-income families where the government pays part of the rent.

After settling into their new home, Marta couldn't wait to start school. On her older sister Maggie's first day, Marta and Juan Luis expected to go with her. But their father told them they were still "too small." They would have to wait.

Finally, when Marta turned five, she started kindergarten. It wasn't quite what she had expected, though. Marta was still learning English, and she felt bewildered most of the time. She realized she was different from the other students in her class and often wished that she were one of the Anglo girls so she would fit in.

While Marta was adjusting to school, her father was working day and night. When he finished one 8-hour shift at the steel mill, he would start another at the Ford Motor Company. Working two jobs—"moonlighting," as it is called—was a dangerous thing to do: If the steelworkers' union found out about it, Toby could have lost his job. But as Toby knew from experience, sometimes risks had to be taken.

Toby's hard work paid off: Often working five 16-hour days in a row, he was able to make a down payment on a house in the

A beaming Marta stands in front of the family's first house *(opposite)*. Maggie *(above left)* and Marta pause for a picture in their new kitchen, wearing dresses their mother made.

11

Taking a break from his 16-hour workday, Toby enjoys time with Azucena and (clockwise) baby Irene, Maggie, Marta and Juan Luis. (Gloria not pictured.)

Detroit suburb of Lincoln Park in 1956. He had also managed to squeeze in night classes to prepare for the test to become a U.S. citizen, which he passed in 1953.

The Tienda family packed up their belongings once more and moved into a bungalow on Hartwick Street, a neighborhood of small, closely spaced houses. Now they were seven: Toby and his wife, Azucena, Maggie (age eight), Marta (six), Juan Luis (five), Irene (two), and Gloria (not yet one).

~ An All-American Mexican American Girl

Like Toby, many of the Tiendas' neighbors worked in the steel mill or in the Ford or General Motors plants nearby. Many also worked two jobs. The Tiendas were the only Mexican American family on their block. But they had a lot in common with their neighbors, most of whom were of Eastern European or Italian descent. All of them believed in the American dream. All of them worked hard to provide a better life for their children.

Siblings Juan Luis, Marta, and Maggie had a very close relationship.

As Marta and her siblings explored their neighborhood, they made new friends. They caught frogs and snakes in the big field beyond the railroad tracks, which ran through the Tiendas' neighborhood just five houses away. The other children were intrigued that the Tiendas spoke Spanish. Often they would urge Marta and her siblings to "say it in Mexican."

12

When September of 1956 rolled around, Marta began first grade at the neighborhood school. The Max Paun Elementary School was a one-story, U-shaped brick building. In the back was a gravel playground with swings and a merry-go-round. Beyond that, a wide-open field beckoned.

In school, the Tienda children were always on their best behavior. They knew that if they misbehaved and were scolded by a teacher, their father would punish them again when they got home. Having struggled so hard to give his children this opportunity to learn, Toby had no tolerance for classroom mischief.

> She still felt like she was different from the other students, and that bothered her.

Marta was a fast learner. By now she had mastered her ABC's and could read English pretty well. Even so, she didn't always understand everything her teacher, Mrs. Blanche, told the class. She still felt like she was different from the other students, and that bothered her.

~ Tragedy Strikes

In the spring of 1957, Marta's mother got sick. After several unhelpful visits to doctors, she was admitted to the hospital for gallbladder surgery.

At school during this time, Marta and her first-grade classmates were making Easter cards to bring home to their parents. They were supposed to trace half of an Easter bunny on a card, cut it out with scissors, then open the card to create a whole rabbit.

Marta thought this would be just the gift to cheer up her mother. But she got flustered while cutting it and ended up with two half-bunnies. Marta felt her face flush red with embarrassment. Looking around, though, she saw she wasn't the only one who had made that mistake. So she started working on another card.

Once again, Marta wound up with two half-rabbits. As she fell behind the rest of her classmates, who had now moved on to decorating their cards, she panicked.

Marta tried one more time. As she carefully cut around the bunny shape, she urged herself, "Don't cut the back, don't cut the back." But she did cut it. Marta quietly started to cry.

Mrs. Blanche walked over to see what was the matter. *She's going to say something to make me feel better,* Marta thought to herself. Instead, Marta got the spanking of her life! With each blow, as the teacher's jangling bracelets clanged in her ear, Marta's humiliation grew.

To the little girl so worried about her mother in the hospital, the bunny card seemed like the one thing she could do to help. Mrs. Blanche's harsh punishment for her innocent mistakes seemed especially cruel under the circumstances. The situation left Marta with a deep-seated fear that she would be punished for any mistakes she made in life.

Marta's maternal grandmother—Grandma Socorro—had arrived from Texas to look after the children while their mother was in the hospital. On April 14, 1957, the Sunday before Easter, Marta's mother died of complications from surgery. Yet no one told Marta and her siblings that their mother had died. Toby and Grandma Socorro were so distraught over the loss that they did not want to subject the children to their overwhelming grief. Without her children's knowledge, Azucena was buried in a quiet ceremony. And the Easter card that Marta had finally finished after toiling over it so diligently? It never reached her.

Marta (in swing) and Maggie enjoy time in the park with their mother several years before her death. The children never had the chance to say goodbye to their mother before she died.

Settled in his favorite easy chair days later, Toby gathered his five children around him. Quietly, hesitantly, he informed them, "Mommy's with God."

Marta was confused. "You mean she died?"

"Yes," Toby replied.

Marta was devastated. Because her mother had already been buried, there was no way for her to say goodbye.

~ Back to Texas

Grandma Socorro took Marta and her four siblings back to Texas so that Toby could keep on working his factory jobs in Detroit. Marta found her new school frightening and strange. The classroom doors opened directly to the outdoors, and her new teacher, Mrs. Jones, was even scarier than Mrs. Blanche. Marta watched in terror as a misbehaving student got paddled right outside the door. Because she couldn't figure out what the student had done to earn the beating, she feared she would be next.

Marta's grandmother, Grandma Socorro, took care of the Tienda children for several months after Azucena's death.

Marta had a tough time understanding the accents of the children in South Texas. Even simple words were pronounced differently there; the, for example, came out as "thee," rather than "thuh." Although Marta feared that her accent would single her out for punishment, her teacher turned out to be far more sympathetic than Mrs. Blanche. Marta never did receive that dreaded paddling outside the open school door.

As the school semester neared an end, things grew tense in her grandparents' house. Marta's step-grandfather constantly grumbled about having five extra mouths to feed. Feeling unwanted and missing their father, the Tienda children wrote a letter to Toby, begging him to bring them home.

~ Other Lands, Other Lives

When summer came, Toby drove the 1,700 miles from Lincoln Park, Michigan, to get his children and take them back up north. After stopping briefly in Texas, he decided to drive another few hundred miles to visit his family in Mexico.

Toby, his two nieces, and all five of his children piled into his maroon 1953 Chevrolet. The two nieces sat up front with Toby, while the kids wedged their bodies onto bedding thrown over suitcases in the backseat. The car had no air conditioning, making the long road trip sizzling hot.

This was Marta's first trip to Mexico—and, it turned out, her only visit ever to the home where her father had grown up. Toby's family lived so far out in the country that the eight travelers had to ride an oxcart the final leg of the journey.

Although she was just nearly seven years old, Marta could see that her father had escaped a life of harsh poverty. All nine members of his family lived in two thatched huts with mud floors, one for sleeping and the other for cooking.

Despite these spartan conditions, Toby's family welcomed their guests warmly and tried to make them as comfortable as possible. They pulled down a mattress stored in the rafters for Marta and her siblings to sleep on. And, in honor of their American relatives, the family killed and roasted a pig for a feast.

~ Fighting off Foster Care

A few days later, the Tiendas returned to Lincoln Park, where the enormity of their loss sank in. Now a single parent, Toby had no choice but to leave his children at home when he went to work. If he couldn't find someone older to look after them, he left 10-year-old Maggie and 8-year-old Marta in charge of the three younger children. A neighbor reported this unsafe arrangement to the local child-protection agency.

One day after Toby got home from work, two officials from the agency knocked on his door. They threatened to take his children away and place them with foster parents.

"Over my dead body," Toby countered.

Seeing Toby's devotion to his children, the workers relented. They gave him time to find an older babysitter for the children.

At first, Toby's two nieces looked after the children. But they returned to Texas when September rolled around, forcing Toby to make do with a string of hired babysitters who stayed but a short time and then moved on.

When Marta started second grade, her family was still adjusting to life without Azucena. Despite the challenges, the Tiendas managed to stay together.

~ Charity and Heartache

For the first and only time in his life, Toby, a proud man, was forced to accept charity. The Tiendas received "surplus commodities"—basics such as peanut butter, dried milk, and canned meat—from the government. The dried milk tasted terrible, so Marta and the others mixed it with regular milk. They also got large bricks of cheese and big bags of flour, which they used to make tortillas.

> The Tienda children felt the loss of their mother more intensely than ever.

The Catholic Church also helped out by donating clothes. Once, a woman from the church took the children to a Sears store to buy them new shoes. Marta's gaze fell on a pair of black patent leather shoes with tiny white stripes. They were so shiny she could see her face in them. Marta had never wanted anything so much in her life.

"Here, dear," the church woman said, "try these on." She held out a pair of clunky saddle shoes—dull white lace-ups with black patches on the sides.

"But I want these shoes," Marta said, pointing to the flashier pair.

"Those are too expensive," said the woman. "Besides, they're not practical."

Marta wound up with the saddle shoes. Her hatred of those shoes sparked a lifelong love of fancy footwear.

Christmas that year was sparse. A nonprofit group, the Goodfellows organization, supplied a small tree for the Tiendas to decorate, as well as a single practical gift for each child. The Tienda children felt the loss of their mother more intensely than ever.

Marta longed for a pair of shiny black patent leather shoes, but ended up with a more practical pair: black and white saddle shoes, like those shown at left.

Toby and Beatrice got married
in November 1959.
At last, it **seemed,**

the children would have

a mother

to look after them.

SHELTER FROM THE STORM

4

A t a local dance in the summer of 1958, Toby met a Mexican woman named Beatrice Smoot. She had grown up in West Virginia, where her family had fled to escape the aftermath of the Mexican Revolution. There, her father worked in the coal mines. Beatrice had a 13-year-old son, Jack, from a previous marriage.

Toby and Beatrice got married in November 1959. At last, it seemed, the children would have a mother to look after them. And because Beatrice worked at the local dairy market (a small grocery store), the family would have more money.

Beatrice, a heavy woman, seemed as nice as she was enormous. She made sure all the girls got fancy new dresses for the wedding— and Marta finally got those patent leather shoes.

At their first Christmas as a new family, the Tiendas had a real Christmas tree with lots of presents beneath it, all wrapped in beautiful paper. Each child received not just one gift, but many! It was one of the best Christmases they had ever had.

Marta, Maggie, Irene, and Gloria (opposite, clockwise from top left) show off their fancy new dresses and patent leather shoes. Their step-mother bought these outfits for the children to wear when she wed their father. At Christmas Marta (above) and her family enjoyed more happy times.

~ The Tomato Pickers

The good times did not last. The next summer, Marta's father lost his job because of a strike in the steel mill. Toby piled the family into their 1957 Ford station wagon and drove them to Monroe

County in eastern Michigan, where they would work in the tomato fields until the strike ended.

This was the first of two summers the Tiendas spent picking crops. While Marta and her brother and sisters went out into the tomato fields every day with their father, Beatrice, now pregnant, stayed at one of the shacks where the workers and their families lived.

They would rise before dawn, the sky streaked with purple, and plod out to the fields with their baskets or sacks. Marta and Maggie usually worked as a team. It was backbreaking work, but they each had to fill their quota to get paid: 10 pecks (about 20 gallons) of tomatoes in the morning and 10 more in the afternoon. They had to take care not to crush the ripe tomatoes; no one would buy those.

Before they can be paid, pickers must line up to have the beans they picked weighed. Marta had to do the same with the tomatoes she gathered.

Out in the fields, Marta would stare at the grower's big white house with its high porch and pillars. What did it look like inside? What sort of people lived there? Were the children just like her, or different in every way? Compared with the shack where Marta and her family stayed, the grower's house seemed like a castle. *You'd have to be really rich to live in a house like that*, thought Marta.

Muddy and exhausted at the end of each long day in the tomato fields, Marta and her family washed up at a communal outdoor pump—the shacks had no running water. Then they used a hot plate to cook dinner—the shacks had no kitchens.

After dinner, they sat around a campfire with the other migrant families, mostly Mexicans, too tired from the day's toil to make much conversation. For some, this was a way of life. For Marta, however, it was just another test to pass before she could return to school.

~ Hardships at Home

As soon as her new baby brother, Reynaldo, was born, Marta's relationship with her stepmother took a nose dive. Beatrice had gained so much weight during pregnancy that she developed diabetes, a disease that demands daily injections of insulin (a hormone drug). She was unable to work because she had difficulty walking. More troubling still, she seemed to show no interest in caring for Reynaldo.

That task fell to the Tienda children, who took turns babysitting. Marta and Maggie now bore some of the same burden as parents. Instead of playing with their friends after school, they rushed home each day to take care of their infant brother.

> Marta's household chores gradually piled up until she felt like Cinderella, buried beneath a mountain of tasks.

Tensions in the tiny house on Hartwick Street mounted steadily. The two-bedroom house had seemed small before. But now, with the addition of Beatrice, Jack, Reynaldo, and sometimes Beatrice's mother, it was bursting at the seams. Every morning, nine people—Marta, Maggie, Juan Luis, Irene, Gloria, Toby, Jack, Beatrice, and her mother—jostled to use the house's sole bathroom.

The only refuge for Marta and her sisters was the attic they shared. At one end of the attic, Marta and Maggie each slept in a twin bed with a dresser between them; at the other end, Irene and Gloria shared a sofa that converted to a bed. This makeshift bedroom under the eaves became the girls' sanctuary; Beatrice almost never climbed the stairs.

~ "Because I Said So"

Marta's household chores gradually piled up until she felt like Cinderella, buried beneath a mountain of tasks, with Beatrice in the role of the wicked stepmother. Every day when Marta and her sisters walked home from school for lunch, they found the sink

Marta had to use an old-fashioned wringer washer similar to the one shown below to do the family's laundry. As in the photo, Marta dried the clothes by hanging them on a clothesline.

overflowing with dirty dishes. The girls had to wash them before they could return to school. After school, more chores awaited before they could start their homework.

For Marta and Maggie, that usually meant washing clothes—and there was always plenty to wash. Marta longed for an automatic washer and dryer, but Beatrice refused to buy one; the old-fashioned, hand-fed wringer washer, she insisted, did a better job. But Marta's aching arms told her a machine would do the best job of all. A clothesline stretched across the basement served as the dryer in the winter months. In the summer, the girls hung the clothes outside to dry.

If the Tienda children wanted to go to a movie on Saturdays, they had to plan their escape well in advance. They would get up early, scrub all the floors in the house, wash the dirty laundry, and carefully iron the clean clothes—even the T-shirts and underwear. Then they would wash their hair with eye-stinging laundry soap—the family couldn't afford shampoo—and stampede out the door before Beatrice got angry about something and broke out la quarta, her bullwhip. She had brought this torturous tool of discipline with her from Mexico, and she didn't hesitate to use it on the kids. Juan Luis seemed to feel the sting of her lash the most.

Beatrice wouldn't let the girls swim in the quarry with the other neighborhood kids, nor would she allow them to wear two-piece swimsuits or short skirts. When Marta asked her stepmother why they couldn't do the things her friends were permitted to, Beatrice dismissed her curtly: "Because I said so."

~Safety in Numbers

The Tienda children formed a united front against the harshness they endured from Beatrice. Marta and Maggie sewed dresses for themselves and their sisters. Marta brushed Gloria's hair; Maggie brushed Irene's. Juan Luis polished everyone's shoes. Maggie and Marta made hot cereal for breakfast.

In Juan Luis, Marta found a "partner in crime." They went to school together, and Marta stuck up for her brother whenever Beatrice mistreated him.

Next door to the Tiendas lived Lucille Page, a woman who frequently sided with the Tienda children in their battles against Beatrice. If Beatrice was the wicked stepmother, Lucille was like a fairy godmother. She and Marta shared a rebellious streak. When Lucille was younger, for example, she had attended college against her parents' wishes. They believed—as many did at that time—that women did not need a college education.

Dressed up for Easter Sunday in 1960 *(left)*, Marta smiles for the camera. As life with her stepmother grew more stressful, Marta turned to her neighbor, Lucille Page *(below)*, for comfort and protection.

Marta felt Lucille understood her. She encouraged her in school and scrutinized every report card that Marta brought home. She listened to Marta's horror stories about her stepmother and often came to her defense. She offered Marta a safe haven against Beatrice's frequent outbursts and threats. Most important of all, Lucille never judged Marta.

More than Marta's protector, Lucille was a model of what girls could hope to become when they grew up. Twenty years before the advent of the personal computer and 30 years before the Internet, Lucille used a computer in her work as an accountant for a company that made plastic products such as hula hoops and traffic cones. Lucille was determined to achieve her goals in life. She had attended college classes at night while working during the day, and she drove a Cadillac—a sure sign of success in the 1960s.

Marta also couldn't help being impressed by Lucille's stylish high heels. Those fancy shoes were more than just a fashion statement; they were evidence that a woman could be smart, successful—and cool.

Marta had taken
to heart her father's relentless
message:

Doing well in school
was the ticket
to a better life.

A Dawning Dream

5

Home life remained turbulent for Marta and her siblings, and they turned to school for a taste of success. Marta and her brother, Juan Luis, had taken to heart their father's relentless message: Doing well in school was the ticket to a better life.

Marta did well in all her subjects. But by seventh grade, math and science gave her a special thrill—the thrill of figuring things out. She liked the idea of facing a problem and finding its solution. She learned about the scientific method: Scientists often start with a question about something, then look for evidence to help them answer that question. She learned the importance of gathering information and facts, known as data, and then analyzing them.

Marta began to find meaning in numbers. The way Marta looked at it, numbers held secrets—and math and science were the keys to unlock those secrets.

That year, one of her school assignments was to bring in a scientific article from a magazine. But the Tienda household didn't subscribe to any magazines, so how could Marta complete the assignment? A neighbor, Mary Fields, let her leaf through a pile of *Life* magazines, where she found an article on hepatitis (a liver disease) and the research under way to cure it. Marta cut out the article and brought it to class. Not only did her teacher like it, she made it the classroom model.

At Huff Junior High School *(above)*, Marta *(opposite)* found pleasure in the challenges of math and science. Teachers noticed her determination to do well and encouraged her in her classes.

~ Making Beehives Behave

Despite Marta's dawning fascination with math and science, the poverty she grew up in limited her view of the career options ahead. As a student in junior high, Marta decided she wanted to be a beautician when she grew up. She loved curling the neighborhood women's hair, and she earned money doing it. One neighbor, Mrs. Moore, paid Marta a quarter to put her hair up in pincurls. When Marta used a sticky green styling gel, Mrs. Moore gave her an extra dime.

> Marta's hair was another story. She had so much of it, and it was so curly, that nobody knew quite what to do with it.

One of Marta's specialties was the beehive, a hairstyle popular in the 1960s. Marta would tease her customer's hair with a comb until it stood up high. Then she would smooth out the top layer and spray it until it was frozen stiff, making the woman look like she was wearing a beehive on her head. Marta loved using her talents to benefit others. Her customers always felt good about themselves after she had worked on their hair.

Marta's hair was another story. She had so much of it, and it was so curly, that nobody knew quite what to do with it. One way to get silky-smooth hair, she learned, was to roll it around juice cans at night, then sleep on it that way. In the morning, Juan Luis would stretch her hair out on the ironing board and iron it flat.

Eighth-grader Marta *(right)* pictured at a family gathering. Marta never missed an opportunity to get dressed up.

~ A Glimpse of the Road Ahead

When Marta wasn't making her neighbors look glamorous, she attended Huff Junior High School. What a change from Max Paun Elementary! The school was huge, with staircases leading up three stories. Finding her classrooms and storing her gear in lockers took some getting used to.

And then there were the older kids—especially the girls. Many of them had come from Catholic schools, where the discipline was rigid. Now, granted the relative freedom of a public junior high school, they ran wild. With their purple eye shadow, tight skirts, and ratted hair, they struck Marta as exotic creatures.

English was one of Marta's favorite classes. Her teacher, Mrs. Miller, wore elegant sweaters and pearl necklaces and had a shock of wavy, sandy-colored hair. Recognizing Marta's smarts and sensitivity, Mrs. Miller treated her with kindness.

Mrs. Miller also saw how hard Marta worked—a trait that put her far ahead of the class. She wanted to encourage Marta to keep challenging herself, so Mrs. Miller set up a nook in the back of the classroom where Marta could do higher-level work. For several

Marta's time at Huff Junior High School *(above)* was marked by an increasing awareness of her own talents, interests, and potential. It was then that she realized attending college was something she could, and should, do.

weeks, Marta sat in the study nook doing her own advanced work while the rest of the class followed the regular curriculum.

One day, Mrs. Miller asked Marta to do some extra work on gerunds—verbs used as nouns, such as running and thinking.

Marta was concentrating so intently that she jumped at the click-clack of Mrs. Miller's approaching heels. As the teacher looked over her work, Marta said a silent prayer that she had done everything correctly. When she glanced up at Mrs. Miller, the teacher had a far-off look in her eyes. *What is she thinking?* Marta wondered. Then Mrs. Miller sat down at an empty desk next to Marta.

Marta didn't realize it at the time, but she was doing sociological work when she conducted a door-to-door survey for a science class in eighth grade. Her research project made it into the local newspaper.

Wednesday, June 3, 1964   The Lincoln Parker,

—The Mellus Newspapers' P

GETTING ANSWERS—Bill Schenk and Mai Tienda, eighth-graders in George Moore's scie classes at Huff Junior High School, Lincoln Park, w among Moore's pupils who conducted a door-to-d survey Thursday and Friday to discover how m residents smoke or chew tobacco. They are shown terviewing Mrs. Roderick McGathen, of Lincoln Pa The survey was made in conjunction with a scie unit on "The Use of Tobacco and Its Effects on Human Body." Some 900 persons were interview

"Marta, what do you want to do when you grow up and finish school?"

"I want to be a beautician," Marta replied.

"A beautician? Why do you say that?"

"Because I'm really good at it," Marta said proudly, "and I like it."

"But don't you want to go to college?" Mrs. Miller asked.

College? The possibility had never occurred to Marta. Her father had always spoken of graduating from high school—something he had never achieved—as a lofty goal in life. But he had never mentioned college as part of his dream for his children. So Marta shook her head and said, "No."

"You're such a smart girl. Why not?"

"College is for rich people," Marta said simply. She wasn't

angry. It was just the way she thought things were. Marta knew she was poor, much poorer than her classmates.

Then Mrs. Miller said something that would change Marta Tienda's life: "You could get a scholarship, you know."

Marta knew that deserving students received scholarships to help them pay for college. But a scholarship for herself? That seemed about as likely to Marta as a flight to Mars. Yet here was her favorite teacher, Mrs. Miller, telling her she had a shot at going to college.

> Marta's dreams of becoming a beautician suddenly seemed small. Instead, she vowed, she would become the first college student in her family.

Marta's dreams of becoming a beautician suddenly seemed small. Instead, she vowed, she would become the first college student in her family. Marta pictured herself wearing a cap and gown, walking across a stage to receive her diploma. She could see her father smiling up at her from the audience, full of pride at her achievement.

Just a few words from a caring teacher had opened Marta's eyes to a world of potential.

She would be the

first one

in her family

to attend college.

6

MARTA'S MISSION

By the time Marta reached Lincoln Park High School in the fall of 1965, she was desperately seeking more independence. She wanted to dress more fashionably. She wanted to go to school dances. She wanted to go out on dates.

Marta's stepmother, however, took a less enlightened view. Beatrice seemed to want Marta to live the isolated life of a teenager in rural Mexico. Her automatic reply to Marta's requests was, "No!"

Once, during her senior year in 1968, Marta was writing a research paper that required a trip to the Wayne State University library, not too far away in Detroit.

"Why do you have to go all the way over there?" Beatrice challenged her. "If the local library is good enough for your brothers and sisters, it should be good enough for you."

But it wasn't good enough for her, and Marta knew it. The local library would never have the kind of information she was seeking.

Indeed, Marta's ambitions seemed boundless on a number of fronts. Though not a natural athlete, Marta was such an eager competitor in team sports that her fellow students elected her president of the more than 200-member Girls' Athletic Association (GAA) in her senior year. Whatever the GAA did—be it a sports banquet or a competition—Marta jumped in to help out. Even though she was not a great player, she often ended up on the winning team when they played touch football, volleyball, softball, or basketball.

Marta's father couldn't have been prouder of his daughter on the day she graduated from high school in June 1968 *(opposite)*. Marta's senior yearbook photo is shown above.

She was admired by her classmates for her leadership and her spirited nature. Her friends were always eager for her to join their team.

She also outworked her peers. After school she would babysit the neighbors' children, iron laundry for $5 a basket, or put in hours as a pharmacy cashier. Then, after getting home from work around 10:30 P.M., Marta would hit the books with a vengeance. Some people called her a brain. But Marta didn't see it that way. She just worked hard, and didn't stop until the job was done.

~ Defying Limits

Marta's hard work and leadership in high school paid off. She was inducted into the National Honor Society *(top, front row, center)* for her grades. Marta *(above, fourth from left)* plays around with fellow members of the Girls' Athletic Association.

Not everyone applauded Marta's drive to succeed. At home, Beatrice resented her stepdaughter's aspirations, and she made sure Marta knew it. "All this talk about college," she taunted Marta, "from a girl who's just going to grow up and get married!" To drive home her point, Beatrice refused to give Marta an inch of leeway. In return, Marta questioned, defied, and openly disobeyed Beatrice.

Marta had watched Maggie's humiliation as she covered up her hairy legs with knee-high socks and long skirts. Marta wasn't about to stand for that. All the other girls at school were shaving their legs—why shouldn't she?

So Marta shaved her legs—an act that Beatrice had forbidden in her household. There was trouble when Beatrice found out, but the deed was done—there was nothing Beatrice could do about it.

On the day that Beatrice threw a flowerpot at Marta, she and her brother decided to run away in the family car. Marta, who had earned her driver's license a year earlier, drove Juan Luis to their church to seek help from their priest. Perhaps he could find them a safe place to live? The priest listened sympathetically, but it was clear he wasn't prepared to help them.

Marta and Juan Luis leaned on each other when life with their stepmother got difficult.

Marta and Juan Luis drove around aimlessly for hours. Then, dejected and hopeless, they headed back home.

Finally made aware of the torment that Beatrice was inflicting on his children, Toby acknowledged his own regret about the situation. "I'm so sorry," he told Marta. "I thought I was getting you a mother. I thought I was doing the right thing."

~ College Bound

In her senior year at Lincoln Park High School, Marta applied to only one college: Michigan State University. Although she had excelled in school, graduating third in a class of nearly 600, her high school guidance counselors did not urge her to approach any other colleges and universities. Most Lincoln Park High School graduates did not go on to college. Her counselors may not have been accustomed to advising a student of such unusual promise.

Marta, Rudee Rossi (her best friend in high school), and their dates are decked out for the senior prom.

Still, when the acceptance letter arrived from Michigan State, Marta felt giddy to be heading for college. A state scholarship for low-income students would cover the tuition, and Marta's savings from working nearly full-time during her senior year and over the summer would help with other costs. She would be the first one in her family to attend college.

~ Natural-Born Bookworm

Marta started classes at the East Lansing campus of Michigan State University in the fall of 1968. She and her best friend from Lincoln Park High School, Rudee Rossi, were roommates that first year, along with two other girls.

Marta and her high school pal, Rudee Rossi *(above)*, roomed together during their first year at Michigan State University.

When they arrived on campus, they found that their dorm room looked as sterile as a hospital ward, so the girls did their best to spruce it up. Rudee had brought her sewing machine with her and the two friends worked together on a tight budget to decorate the place. They covered their bunk beds with a green canopy, added matching bedspreads, and hid the bare tile floor beneath shaggy throw rugs of green and turquoise. Their funky, comfy sitting area was the talk of the dorm.

Rudee went home to Lincoln Park on weekends, but Marta avoided the trip. For one thing, she and her stepmother battled constantly when she was home. The more independent Marta became outside the house, the angrier Beatrice got inside it. For another, Marta was too busy studying. It wasn't enough to do well—Marta was driven to excel. When she had completed her assigned work, she often sought out and read the materials her professors had only mentioned in passing. Her intense desire to succeed drove her to study hard, and she earned mostly A's during her freshman and sophomore years.

~ A Dose of Reality

Marta began her college education with a vivid dream: One day she would return to Lincoln Park High School to teach Spanish. She wanted to improve upon the performance of the Spanish teachers she had encountered during her own high school career. They may have understood the language, but their lack of passion for it made it clear to Marta they did not feel it in their bones.

Marta began to pursue her dream by taking courses in Spanish literature, along with classes that would lead to a teaching certificate. All would-be teachers visit schools to watch other teachers on the job; they then teach a semester themselves as a student teacher. This gives the future teacher classroom experience. It also tells her or him whether teaching is the right career choice.

In the spring of her junior year at Michigan State, Marta taught Spanish at a small suburban high school in Charlotte, Michigan, teaching ninth graders and seniors. After starting off with great enthusiasm, she became troubled by what she saw. Some of the teachers seemed disillusioned by their

Marta *(center, blue shirt)* with fellow students in one of her high school Spanish classes. Her classroom experiences—good and bad—inspired her when she started teaching Spanish.

jobs, giving up on the particularly hard-to-teach students. Others had resorted to having students memorize lists of vocabulary words in alphabetical order, rather than using the words in conversation. *That isn't teaching*, Marta thought angrily.

She recalled a more positive example: her own tenth-grade Spanish teacher, Mrs. Enedy. Even though Marta spoke Spanish, she hadn't known how to write it, and her vocabulary was limited. Mrs. Enedy did much more than cover the basics; she immersed the class in Spanish culture and taught her students to use the language in context. She encouraged her students to be expressive, to take chances.

Now, at the high school where Marta was student teaching, only one teacher showed such creativity. To Marta's dismay, that teacher was fired for not following the curriculum more closely. *Fired?* Marta thought. *For being a creative teacher?* She wanted no part of a system that tolerated poor teaching. Marta's own negative experiences as a student teacher made her re-examine her fantasy of returning to Lincoln Park to teach Spanish. Maybe that wasn't the right career path for her after all. But then, what was?

Marta didn't have an answer to that question, but she did know one thing for sure: Whatever she chose to do, Marta desperately needed to feel that she was making a difference in people's lives.

Marta was

troubled

to discover

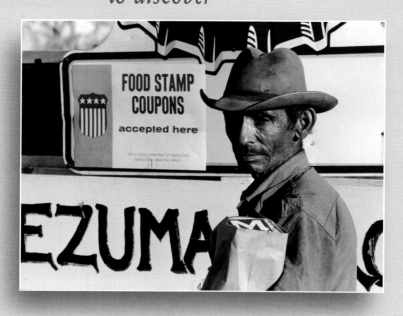

this
"us against them"

attitude.

7 7

BONUS DAYS

"C ongratulations!" said the voice at the other end of the
phone line when Marta picked it up near the end of her
junior year in college. "We've accepted your application
to work for Michigan's State Department of Agriculture!"

Marta couldn't quite remember applying for this particular job,
but she decided to return the enthusiasm. "That's great!" she said.
"When do I start?"

Marta's spontaneous energy was rewarded. The job—in Alpena,
northern Michigan, about 200 miles from her college campus in
East Lansing—seemed like it would be interesting work. It also
meant Marta could live on her own, escaping another summer
with Beatrice.

Shortly after classes ended in May 1971, Marta left for Alpena
with her friend Rudee. Rudee stayed the weekend and helped
Marta settle into a small studio apartment. As they had with
their dorm room at Michigan State, Marta and Rudee made the
apartment look homey.

The following Monday, Marta reported for work at the county
government building in Alpena. There, she learned that she'd be
certifying migrant workers for food stamps.

Sometimes as Marta was interviewing the workers and helping
them with their paperwork, her boss, the director of county social
services, would drop by. He respected the migrant workers.

During her summer
job in 1971, Marta
certified migrant
workers for food
stamps. People used
the coupons to
purchase food for
their families (oppo-
site and above). The
workers Marta met
reminded her of her
own family.

A family returns home, arms full of groceries purchased with food stamps. Marta's work in Alpena, Michigan, helped low income families get the help they needed to put food on the table.

He knew how hard they worked. He would often greet them by heartily crying out, "Bonus days!" That brought smiles to many of the Mexicans, who were too polite to tell him how thoroughly he had just mangled the Spanish greeting buenos días (good morning).

Actually, Marta thought, my boss is right on the money: For these workers, getting food stamps certainly is a "bonus day."

~ Us Against Them

During the Fourth of July weekend, many migrant workers drove their pickup trucks into the town of Alpena to take part in the celebrations. After a hectic week of work, Marta was looking forward to relaxing and enjoying the festivities too. It didn't take her long to discover that some local residents resented the migrants.

One woman, acting friendly at first, asked Marta what kind of work she was doing.

"I'm certifying migrants for food stamps," she replied.

The woman's face scrunched up like she had just bitten into a lemon. "Those people come here and get free housing and food stamps," she groused. "We don't get those things."

Another resident overheard the conversation and chimed in. "The migrants show up in big, fancy pickup trucks," she complained. "We can't afford cars like that."

"You don't understand," Marta responded. "They have to live all year long on the money they make in the summer. Sure, they buy good trucks—but they need those to transport their families from state to state, from job to job. Without the trucks, they couldn't work."

Marta's answers rose directly from her own, and her family's, work experiences. She would have been happy to tell the residents of Alpena what "free worker housing" was really like, for she had lived in those crowded, filthy shacks herself. Marta was troubled to discover this "us against them" attitude.

Workers pick strawberries in a field in Michigan. Having worked in the fields herself, Marta was able to describe firsthand the hardships the pickers had to endure, including the back-breaking work.

~ When Rules Go Wrong

Two weeks into her job, the hectic pace lessened and Marta found herself with some time on her hands. She got in her blue Ford sedan with the yellow Michigan state seal on the doors and drove the flat, winding roads that threaded the berry fields. She wanted to visit the fruit pickers and growers to learn more about their lives.

Migrant workers stand in front of their "free" housing. Facilities such as these were not always well maintained.

To Marta's surprise, some growers invited the college girl into their homes. The houses were hardly the castles she had imagined them to be as a young girl toiling in the fields. Though the growers were much better off than the migrant workers they employed, Marta could see that most of them were far from wealthy. Their household furnishings were modest, and they repeatedly expressed their own fears of making ends meet.

Not only that, but new federal and state laws were making it harder for the growers to hire migrant workers. These rules had an honest purpose: to improve the lives of the workers, many of whom spoke little English and were therefore powerless to protest their living and working conditions. Their housing was often crowded and wretched.

> As Marta talked to the growers and workers, she wondered about ways to meet everyone's needs.

The new rules required a minimum amount of space for each member of a worker's family. If a worker showed up with more family members than the government said the grower's housing could hold, the grower could not hire that worker.

As a result of these new rules, some workers were being turned away from farms where they had picked fruit or vegetables for years. And some growers, desperate to get their crops picked, couldn't hire enough workers. They lacked proper housing for them.

The growers needed workers to pick their crops. Some crops, such as strawberries, have a short harvest season and must be picked by hand. The workers came a long way to pick those berries and earn money for the rest of the year. If the rules designed to protect them kept them from working instead, they would be worse off than before, not better.

As Marta talked to the growers and workers, she wondered about ways to meet everyone's needs. One day she dropped in on a pickle grower. He had planted 80 acres of pickling cucumbers but could find no one to pick the crop. His wife sat at the kitchen table with her face in her hands, crying tears of frustration and despair.

The workers and the growers, Marta realized, were not "us" against "them"; they were both "we." The workers needed jobs to feed their families, and the growers needed the workers to harvest their crops.

~ "Everybody's Voice Must Be Heard"

Jobs and housing were not the only problems Marta uncovered that summer. As parents, the workers also struggled to find day care for their youngest children. Their only choice was to send their babies to school with their older brothers and sisters—some as young as eight years old.

This put the director of the county migrant education program in a bind. "What do you want me to do?" he asked Marta. "I have to turn the kids away. We can't handle them."

"You shouldn't turn them away," Marta responded.

"But we don't have money for diapers," he said. "And the school doesn't have permission to run a day-care center. We could get in trouble with the authorities."

"We have to get you what you need," Marta told him. "Everybody's voice must be heard."

A migrant worker poses with her child. Marta discovered that parents in the migrant worker camps had trouble finding day care for their young children.

That notion was noble but naive. Indeed, the more Marta thought about it, the more she wondered what, if anything, she could do. *After all,* she told herself, *I'm just a 20-year-old college student working a summer job.*

But Marta was also a young woman whose hard life had given her a steely resolve. She organized a meeting where everyone—workers, growers, officials from the state capital in Lansing, her bosses—could air their gripes. Thanks to Marta's leadership, these people not only started talking about their problems, they also began trying to solve them.

One concrete result: The school received permission to set up a day-care center.

~ *Problem Solver*

Getting the day-care center established was a good deed. It also made news: On July 24, 1971, Marta's picture appeared in the local newspaper, the *Alpena News,* touting her success. A follow-up article stated that the Department of Social Services would accept cribs and playpens for the day-care program and that donors should contact Marta.

The *Alpena News* features Marta and two of her supervisors in an article about the successes of the town's Migrant Program.

PROGRAM SUPERVISORS — Bill Romstadt, director of the Migrant Program in the Alpena area for four years, looks over a county map with Phil Mesa, who recruits children for the educational portion of the program, and Marta Tienda, Department of Social Services migrant family worker. One of the aims of the program is to seek out the migrant children in the community and encourage them to attend classes which are held at Green School during the summer.

That summer had opened Marta's eyes—and mind. She liked being a problem solver—someone whose actions improved people's lives. Marta didn't need any more convincing: Teaching Spanish was not the way she wanted to spend the rest of her life.

Just what career might be calling her, however, remained a mystery.

"If being Mexican means I'm brown, then I'm brown.

But what does it matter?"

ENTER THE MAESTRO

8

When Marta returned to Michigan State in the fall, her Spanish professor, Frank Pino, encouraged her to apply for a special Ford Foundation scholarship. The scholarship had been set up to encourage Hispanic students to seek higher education, but it made Marta uneasy. Never before had she been given an opportunity simply because she was a member of a minority group. She didn't want to receive preferred treatment because of an ethnic label. She wanted to succeed because of her hard work, good grades, and job skills.

Marta recalled the moment in college when she first became aware of the power of these ethnic labels. In the winter of 1972, she was with her brother, Juan Luis, who was a freshman at Michigan State. He introduced her to some of his Hispanic friends, who were trying to recruit students for the college's new Chicano Studies Program. The students kept using the word "brown" to describe themselves.

"Why do you always use that word?" Marta asked them.

"You're brown, Marty," one young woman replied. She had long black hair and light brown skin.

Marta bristled. First, she disliked being called "Marty." Second, she resisted being put in a category. "You're absolutely wrong," she replied. "I'm not brown, I'm white. Can't you see?"

While at Michigan State University, Marta is inducted into the prestigious Mortar Board Society *(opposite)*. Marta clowns around with fellow inductees *(above)* for a local newspaper photographer.

She thought the term "white" referred to the color of her skin. She held out her arm for the woman to see how light she was.

"You're not white," the woman shot back. "You're Mexican."

Marta had always been proud of her heritage. Until then, she didn't see it as a label that defined her. "Okay," she relented. "If being Mexican means I'm brown, then I'm brown. But what does it matter?"

Now, pondering this exchange a year later, Marta realized that the Ford Foundation scholarship wasn't just a "reward" for being a minority. It was a chance for her to plant some seeds for her future. The hard work—nourishing those seeds and making them grow—would be up to her. Marta spent a weekend filling out the seemingly endless application.

~ Good News, Bad News

While Marta waited to hear back from the Ford Foundation, she was busy finishing her last quarter at Michigan State University. Nearing another transition in her life, she looked forward to the changes ahead. What she didn't realize was how much her family's life was about to change as well.

In January 1972, Beatrice demanded a divorce from Toby. Marta's father was stunned. The Tienda children were relieved to have Beatrice out of their lives, but that relief came at a high cost.

> Marta had always been proud of her heritage. Until then, she didn't see it as a label that defined her.

The home Toby had worked so hard to purchase nearly 20 years ago was owned by both Beatrice and Toby. When the property settlement for the divorce was finalized, the courts awarded the house on Hartwick Street to Beatrice. Toby was awarded Beatrice's old house, a run-down shack in Detroit. Although devastated to be losing the home they grew up in, Marta's family pulled together and pooled their resources to fix up the Detroit house. At least Beatrice would no longer be there to torment them.

~ Broken Tortillas

Marta finished her studies and received her bachelor's degree from Michigan State University early, in March of 1972. She had been offered a position teaching high school Spanish, but she decided instead to accept a job with the Michigan State University Cooperative. This was an outreach program created to help educate the citizens of Michigan about practical issues (such as home economics) that could help improve their lives.

Marta worked for an outreach program following her college graduation. Her job was to advise people on how to make their diets more nutritious.

When Marta was hired, the state wanted to do a better job of helping Michigan's growing Hispanic population. Marta visited poor Hispanic communities around the state, including migrant worker camps, to advise people on how to improve their nutrition.

Marta didn't have any background in nutrition. But the program gave her instructions on what advice to offer people to enrich their diets. "Try putting dried milk into the mix for tortillas," she told them. "That will give you more calcium, which builds strong bones."

"We tried that," they told her, "but the tortillas break!"

This kind of teaching, Marta quickly saw, was a two-way street. She knew little about nutrition—how could she do her job? Maybe it would be better to teach something she knew a lot about: sewing. All those hours Marta had spent years ago sewing outfits for herself and her sisters were about to come in handy.

The next time she visited the migrant worker camps, Marta asked the people there, "Would you like to learn to sew?" The answer was an avid "Yes!"

Marta went to Singer, a company that makes sewing machines, and asked them for a donation. They gave her two machines, which Marta lugged from camp to camp, teaching the girls to sew.

~ Decision Time

Marta had finished only one week on the job when a letter arrived from the Ford Foundation. She had won the fellowship to attend graduate school! Not only would the foundation pay for her tuition and textbooks, but it would give her $250 a month to live on.

Marta was torn. Should she accept the fellowship? She wasn't sure she wanted to continue studying Spanish literature, and $3,000 per year wasn't much to live on. She also felt guilty about quitting her new job so soon after she had started.

Marta's first sociology professor, Harley Browning, became a huge influence on her life and career.

Marta turned to the one person who had always helped her in the past: Lucille Page, her next-door neighbor from Hartwick Street in Lincoln Park. After listening to Marta's plight, Lucille told her, "You've got to get your education now. Take the fellowship."

~ Papercitos

Marta resigned from her job, then headed for the University of Texas at Austin, which had the top Spanish department at the time. She began taking courses in Latin American studies, learning about the people of Central and South America.

On a whim, Marta enrolled in an undergraduate course on modern Mexican society, taught by a sociologist named Harley Browning. It proved to be one of the luckiest moves she ever made.

The books Dr. Browning assigned were filled with numbers—charts and graphs and tables. These were the tools of the social scientist, and they told a story about the people who were being studied. When Marta learned that sociologists could use those numbers to make a difference in people's lives, she found herself growing intrigued by this unfamiliar field.

Dr. Browning, for his part, was impressed by Marta's eagerness. One of the first professionals to see Marta's potential as a sociologist, he quickly emerged as her mentor. "I was always on the lookout for good students for our program," Dr. Browning reflects today, "and I could see that she was not only bright but motivated. Her tremendous energy and enthusiasm impressed me from the beginning. I'd never met someone who went 150 miles per hour like that."

Marta had studied literature in college. Now, as she began to learn about sociology, she saw a connection between the two subjects. Novels and poems are art forms that tell stories about people, holding a mirror up to the times in which they are written. Marta came to see the study of people—that is, sociology—as her art. By pinpointing groups in need, she realized, sociology could build bridges to a better future for us all.

In one paper that she wrote about poor communities in developing countries, Marta imagined what she would do if she were a repairwoman who could fix entire nations. In another, about the national character of Mexico, Marta invented an imaginary dialogue between two real people, Mexican poet Octavio Paz and American feminist Gloria Steinem.

> When Marta learned that sociologists could use those numbers to make a difference in people's lives, she found herself growing intrigued by this unfamiliar field.

Dr. Browning was struck by Marta's energy and originality in these and other short papers he assigned. He called them papercitos— "little papers."

On Marta's first papercito, he wrote a few encouraging comments.

On her second, he asked what she planned to do in graduate school.

On her third he wrote, "I'm most impressed by your work. You've got a real talent, a clearly great capacity for hard work, and a bright-eyed enthusiasm for ideas. I'd be most happy, and honored, to sponsor you for entry into the department of sociology. We need you."

That third note decided the course of Marta Tienda's scientific career.

~ Getting Up to Speed

Marta's path was clear at last, but she lacked the tools to navigate it. She hadn't taken courses in two basic subjects—statistics and research methods—that every sociologist needs. Statistics is the branch of mathematics that studies data. Often these numbers come from experiments and surveys such as population counts, so making sense of them is crucial to social scientists.

Marta enrolled in a graduate-level statistics class. Her first day of class taught her that she had overshot her ambition: She didn't even know the basic principles of statistics, such as the meaning of a standard deviation. *(See box, page 51.)*

Marta needed to learn the basics, and that meant starting at the beginning. She took a freshman-level statistics class and did so well that her instructor asked her to tutor some of the other students in the class. That was her springboard to more advanced classes, where she gained skills in statistics that she still uses today.

Of all Dr. Browning's graduate students, Marta worked the hardest. It didn't matter that she was a graduate student; if she had to take an introductory class in sociology or economics, as she had in statistics, that's what she did. That was vintage Marta: Throughout her career, her drive to learn has been so strong that she has never felt embarrassed not to know something.

Never far from her thoughts were the memories of the poverty she grew up in: the crowded bedrooms, the chapped hands from picking cherries or tomatoes, those clunky saddle shoes. Her dread of returning to that life made Marta bear down on her studies. Long after other students had collapsed into their beds, spent from studying, Marta's desk lamp could be seen burning late into the night.

~ Learning from the Maestro

In English, the word *maestro* means master, especially of an art such as music. In Spanish, the word means teacher—but with the added sense of respect for a master of the art of teaching.

What Is a Standard Deviation?

A standard deviation is a number that describes the spread of data. Imagine two classrooms, each with 30 students. The students take a test and in both classrooms the average score on the test is 75. That means that the students in each class are performing at the same level, doesn't it? Not necessarily.

Both of these graphs show classrooms in which the average score on the test was 75. But the data is spread out very differently. In Classroom A, the scores are clustered closely around 75. The standard deviation, shown in red, is only 2. In other words, the students are performing at about the same ability level.

In Classroom B, test grades varied, or deviated, widely. The standard deviation for that classroom, shown in red, is 5. This higher number shows that students in Classroom B are performing at many different ability levels. The blue-shaded areas on the graphs also show that many more Classroom B students than Classroom A students scored below 70.

Why is knowing standard deviation important? Because sometimes averages hide important information about a

situation that is being measured by numbers. And knowing how data is spread helps decision-making. For example, if the school wanted to assign a teacher's assistant to one of these two classrooms, they might decide that Classroom B would benefit the most from having extra help. Classrooms in which students have varying ability levels are a greater challenge for a teacher.

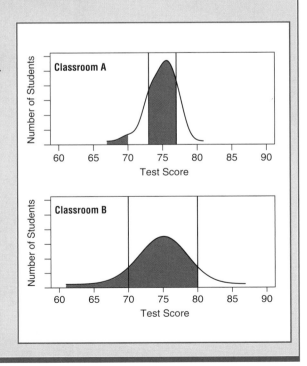

Marta called her sociology professor Harley Browning "maestro." He was a tall man with a full head of wavy white hair, and a distinctive gap between his two front teeth. His office usually looked like it had just been trashed by a hurricane; piles of books and papers, most of them toppled over, lay scattered everywhere. But Maestro Browning always seemed to know just where everything was.

Dr. Browning challenged Marta in every area, constantly pushing her to excel. He fervently wanted Marta to become an expert sociologist. That meant she had to learn to think critically, write forcefully, and ask the right questions.

Professor Harley Browning *(second from left)* reviews the results of a research project with students from the Population Research Center at the University of Texas.

For every class with Dr. Browning, Marta had to write a research proposal. That's a short paper where a researcher poses a question or states a problem, then proposes a possible explanation—a hypothesis—and a way to test that explanation. When it came to making their questions clear and testable, Dr. Browning would not let his students off the hook.

Some questions—"Is there life after death?" for example—are clear but cannot be tested scientifically. That's why social scientists thrive on finding testable questions and coming up with new ways to answer them. Here are a few examples of testable sociology-based questions:

- Does having a baby before age 21 raise or lower a woman's future earnings?

- Do the earnings of immigrants ever catch up with the earnings of natives?

- Does a criminal record make it harder to get a job?

Though a person might have an idea of the answer—a theory—it must be proven scientifically. There's no room for sloppy testing or unfounded opinions. Good answers require good data—numbers gathered from scientific study. Yet data cannot speak for themselves; they give up their secrets only through careful interpretation.

~ Working Women

Energized by her first encounters with the general science of sociology, Marta began to study a specific field—demography— at the Population Research Center, University of Texas, in 1973. Demography is the science of human populations—their size and distribution, how they grow, how they move around.

What Does a Demographer Do?

To better understand the makeup of the American population, many demographers, including Marta Tienda, use the data collected every 10 years by the U.S. Census Bureau. Census data describes how many Americans there are, what racial or ethnic backgrounds they come from, where they reside, the kinds of houses they live in, how many children they have, where they work, how much money they make, and many other details.

Demographers use this census information to figure out how one part of a person's life, such as how much education she has, relates to another part, such as how much money she makes.

With the help of demographers, government agencies and other organizations can use these numbers to make decisions that will improve Americans' lives. A demographer might help a county school board, for example, decide which communities need new schools because they have lots of children. Or a demographer might help local governments decide whether they need to build retirement centers for the elderly.

Demographers also help us understand how the United States compares with other countries in education, health care, crime, and employment. This information helps planners in this country identify what resources are needed where.

The Texas center was a leader in the demography of people from Mexico to Argentina. Researchers there studied topics such as urbanization (the movement of people, usually the poor, to cities in search of better jobs); public health; migration (the movement of groups) within and between countries; and labor markets (who works, and at what).

Marta began her training with a project on Mexican women in the labor force. To find out how working women had affected the economy of five Mexican states, Marta studied how many women worked in each state, and how old they were when they started their jobs. As countries develop—becoming more industrial and less agricultural—women tend to work outside the home more.

Marta's data came from the Mexican census, a nationwide population count done every 10 years. She looked at how many women of certain ages worked outside the home. Her comparison included Mexico's most developed state, the Federal District (which includes Mexico's capital, Mexico City), and its least developed state, Chiapas.

This map of Mexico identifies some of its major cities. Mexico City, located in central Mexico, is part of the Federal District and is one of the world's most populated cities. Tuxtla Gutierrez, the capital of the state of Chiapas, is located in the southeast part of the country.

Proud of the work she had done, Marta presented her research paper to Dr. Browning—and was shocked at his response.

"This is so poorly written that I can't read it," he told her. "You're wandering all over the map. If you want to be a sociologist, you need to learn to write clearly and forcefully. A well-written sentence is worth a thousand pictures."

The maestro's words stung Marta. Fired up, she went back to work on the paper, more determined than ever. She wrote and rewrote, and then she wrote some more. Finally she felt the paper had been fixed.

Her work paid off. The revised paper became the basis for her master's thesis—a lengthy writing project, based on original research that graduate students must complete in order to earn a degree.

~ Feeling the Pinch

Marta had completed her first year in graduate school. She was thriving academically, but otherwise she was practically starving. It was 1974, and inflation was pushing the cost of basic necessities through the roof. Marta's monthly living allowance from the Ford Foundation had been increased to $300. But that was still not enough to meet the skyrocketing price of gas, food, and other living expenses.

At a social gathering, Marta told her economics professor she would have to drop out of the program. The professor tipped off Dr. Browning. Faced with losing his star student, the maestro arranged for Marta to receive an additional $100 per month from the Population Research Center. The added dollars were enough to make it possible for Marta to stay in the program.

Marta hangs out with friends from the Population Research Center. Although she was a focused student, she managed to find time now and then to relax and enjoy herself.

~ I'm Out of Here!

In the summer of 1974, Marta traveled to Cuernavaca [kware-nuh-VAH-ka], Mexico, for a two-month seminar studying women's roles in Latin America. Three groups of students were studying there: some Latin American women, an equal number of non-Hispanic North American women, and a much smaller group that Marta called "tweeners"—U.S.–born Hispanics (Americans of Mexican or Puerto Rican descent).

Very quickly, Marta and the other tweeners felt a gulf open between the Latin Americans and the North Americans. Each group had a different view of the problems facing Latin American women. The North Americans saw one major cause—Latin American men—behind the inequality of Latin American women. These men, they felt, did not see women as their equals; instead, the men believed women should stay home and have children. In the United States, the desire for equality caused women to start the feminist movement.

> Marta did not want the fact that she was a woman—or a Mexican American, for that matter—to influence what she studied or how she studied it.

But the seminar's Latin American women didn't see things that way. The problems women faced in their countries, they believed, stemmed from class distinctions between rich and poor, not battles between men and women.

As tensions mounted between the two groups, Marta grew uncomfortable with the name calling and stereotyping. The North American women acted as if they knew what was best for their Latin American counterparts. That embarrassed Marta, who rejected the view that men were the source of women's problems.

Marta did not want the fact that she was a woman—or a Mexican American, for that matter—to influence what she studied or how she studied it. Upon returning to Texas, a frustrated Marta marched straight into Harley Browning's office. "I don't want to have anything to do with women, or minorities, or Mexicans," she burst out. "In fact, I resign!"

Dr. Browning looked at her calmly. "Is that forever?"

"I don't know about forever," Marta replied, starting to cool off. "But it is for right now."

After her experience in Mexico, Marta worried that she would be pigeonholed as a Mexican American sociologist. If she studied only those people like her, she feared, others might think that was all she could do. She might end up limited to studying Mexican Americans, teaching Mexican Americans, and interacting mainly with Mexican American colleagues.

To be taken seriously as a sociologist, Marta knew, she had to sharpen her research skills on a more general topic. Marta Tienda set out to prove that she was much more than just what she appeared to be on the outside.

What should have been the
happiest day
of her life

became a day of
grief and **pain**.

9

FAMILY MATTERS

I t didn't take Marta long to find her next project. She thought
back to the summer of 1972, when she had helped migrant
workers qualify for food stamps in Michigan. She remembered
seeing women with a lot of children and wondering why—when
they were struggling so mightily to make ends meet—they had
such large families: Wasn't life harder with so many mouths
to feed?

Now, in 1975, Marta set out to answer that question.

Demographers have long noticed that poor countries often
have high rates of population growth. As the countries climb out
of poverty, however, their growth rates usually slow down. But why?
Which is the cause and which is the effect?

Some researchers said rapid population growth creates poverty.
This growth, they reasoned, stems from two factors: a higher birth
rate (more babies are born) and a lower death rate (fewer babies
die, and people live longer, thanks to improved health care). This
leads to more people—both young and old—in the population.
But babies and, usually, the elderly don't work, so they don't
make money; instead, it costs money to support them. This line
of reasoning led some people to argue for a policy of encouraging
people to have fewer babies.

Marta had her doubts.

Marta's wedding to
Wence Lanz *(opposite)*
was a solemn day. The
day before, on August
19, 1976, her brother
Juan Luis *(above)* was
killed in a car accident.

She agreed that many more babies might cost a society—it would have to pay for their health care and education, for example. But might all those extra births benefit a society, too? And who actually pays most of the cost of additional babies: society or the family?

~ Painting by Numbers

Marta's project had started with a question: Are people poor because they have too many babies? Her next step was to restate that question in more scientific terms: Does population growth cause poverty, or does poverty cause overpopulation?

To get her answer, Marta looked at a government survey from Peru, a South American country whose population was booming. Social scientists often use data collected by the government because it is so expensive to conduct surveys on their own.

Next, Marta began to analyze the data. In general, she wanted to see how the family itself changed over time. In particular, she wanted to see what determines whether children work to help support the family. To find out, she compared families that had a lot of kids to those with fewer kids. She also compared families with older kids to ones with younger kids. And she looked at rural and urban families, at families with one parent and those with two, and at poor families and rich.

Marta fed all of this data into her computer and let the machine crunch them into tables. Like any good sociologist, she then closely examined the resulting data tables to interpret them. In essence, she was using numbers to paint a portrait of a people—and to understand how families made decisions.

Marta's analysis revealed that families expanded or shrank in response to changing conditions. For example, a family might add

A Peruvian family assembles outside their farmhouse. Marta studied such families to learn whether poverty and family size are related.

60

adults—usually relatives—to help care for newborn babies or younger children, freeing both parents to work. This tended to increase crowding in the home, yet it also increased support for a growing family.

Also, when economic conditions were poor, families needed more members to handle chores inside the house and to work for pay outside it.

Wow, thought Marta, this sounds familiar.

She flashed back to those Michigan summers when her father had been laid off by the steel mill, forcing the entire Tienda family out into the fields to pick crops. Just like the Peruvian families Marta was studying, her own family had spread the work among the adults and children to make ends meet.

Marta concluded that population growth in Peru did not cause poverty. Instead, the family acted as a social unit, expanding and contracting to adjust to circumstances. This research became her Ph.D. project—one of the cornerstones of Marta's work as a sociologist.

~ Of Love and Loss

In graduate school in Texas, Marta met a student from Venezuela named Wence [WEN-say] Lanz. He asked her to translate one of his papers into English, and they began dating. Until then, Marta had never really dated— she was too busy studying. And despite her feisty nature, she was shy around boys.

The secure, outgoing Wence impressed Marta.

> Just like the Peruvian families Marta was studying, her own family had spread the work among the adults and children to make ends meet.

A generous soul, he seemed to get along well with everyone. But what Marta liked most about Wence was that he was the first person who had ever sat down and truly gotten to know her. Wence listened to Marta. After dating for nearly two years, they decided to get married.

The wedding would take place in Texas on August 20, 1976, after Marta finished graduate school and before she started her first job as an assistant professor at the University of Wisconsin.

Wence's family arrived in Austin from Venezuela to join in the celebration. Marta's family came down from Michigan; her brother, Juan Luis, drove a group of relatives that included their father and their sister Gloria.

This is the last photo taken of Marta's brother, Juan Luis, shown here with his fiancée. He died three days later.

The day before the wedding—and just a day after the exhausting trip from Michigan—Juan Luis borrowed Marta's car to pick up his own fiancée in Houston, a two-hour drive from Austin.

He never got there.

Just 40 minutes outside Austin, Juan Luis's car crossed two lanes of oncoming traffic and was broadsided by a truck. Juan Luis, who police believe had fallen asleep, died instantly.

Marta was devastated. Juan Luis, her brother and best friend, was gone. What should have been the happiest day of her life became a day of grief and pain. But postponing the wedding, her father insisted, would not bring back Juan Luis. And both families had traveled far to witness Marta and Wence exchange vows.

The wedding went on as planned, but without music—and without the joy that Marta Tienda had so looked forward to sharing with her brother.

Right after the wedding, Marta's family returned to Michigan in Toby's Chevy Suburban. Two days later, Marta and Wence rented a truck and followed them north, arriving just in time for the funeral for Juan Luis in Michigan.

~ Remembering Juan Luis

Juan Luis's death hit Marta hard. Just a year apart in age, Marta and Juan Luis had shared a zest for life. Neither would accept the hand they were dealt. Both worked hard to sculpt their own future, refusing to let others shape it for them.

There was one big difference between the two, however: Juan Luis loved adventurous outdoor activities. To him, life equaled excitement. He taught himself to ski. He hunted and camped. Marta enjoyed watching her brother charge through life. Later on, she wondered if he had sensed his life would be short.

Juan Luis had finished high school the year after Marta. He too wanted to go to college, but he had no way to pay for it. So he joined the army, which would pay for him to attend college after his military service. Enlisting in the army wasn't unusual, but enlisting at the height of the Vietnam War—as Juan Luis did—was.

On his way to Vietnam, an officer came to talk to Juan Luis's unit. "Does anyone here know how to type?" the officer called out. Instantly Juan Luis's hand shot up. While the rest of his unit went to Vietnam, Juan Luis went to Japan as a clerk-typist.

He did not, in fact, know how to type—but he learned in a hurry.

After serving in the army, Juan Luis attended Michigan State University, Marta's alma mater. He finished in just three years, then went to law school at the University of Michigan. Because Juan Luis, like Marta, wanted to make a difference in people's lives, he became an activist for recruiting and hiring Hispanic students and faculty. Just before driving down to Texas for Marta's wedding, he had been providing legal help to migrant workers in Michigan.

Juan Luis loved outdoor activities like fishing and hunting. He enlisted in the army (top) during the height of the Vietnam War.

After her brother's death, Marta found comfort in the knowledge that Juan Luis had already improved so many lives. She knew she must carry his banner—but how?

~ Turning to a Familiar Page

The day after burying Juan Luis, Marta and Wence prepared to embark on their new lives together in Madison, Wisconsin, where Marta would soon begin teaching sociology to undergraduate students at the University of Wisconsin. The couple was practically broke: Marta's car had been totaled in Juan Luis's fatal accident, and she and Wence had no savings. In distress, Marta called on the one person who had always come to her aid in the past: Lucille Page.

"I don't know what I'm going to do," a tearful Marta told Lucille. "I've been living in poverty so long, I can't stand it anymore."

Lucille gave Marta $1,000 to help her get started. When Marta began promising to pay it back, Lucille stopped her.

"Don't pay me back," she said. "Just pass it on."

~ Bearing the Banner

Marta poses for a photo just before leaving for an interview at the University of Wisconsin, where she got her first job teaching sociology. She soon became involved with cutting-edge research into the growing Hispanic population in the United States.

Not long after Marta started teaching at the University of Wisconsin, a group of social scientists asked her to be an adviser on the first-ever national survey of Mexican Americans. She would review the questions they planned to ask to help ensure they were worded correctly and could be tested.

Marta thought back to her vow to fight the label "Mexican American sociologist." But enough had changed in her life

64

that she was now ready to embrace the work. For one thing, she had firmly established her credentials as a researcher apart from her ethnic status. Then she thought about another vow— her promise to continue Juan Luis's work in helping Hispanic Americans excel. This national survey would be a first step down that road.

Based on her involvement with the Mexican American survey, the U.S. Department of Labor awarded Marta a grant to analyze the results of the largest survey ever conducted of the nation's growing Hispanic population. Previous studies had looked at Hispanics either as separate groups (Mexicans, Puerto Ricans, Cubans) or as residing in different regions of the United States.

Marta compared all Hispanic groups for the first time on a national basis. She did this by examining the data tables generated by a computer. The numbers helped her compare average levels of education, differences in work activity, poverty rates, income levels, and household composition (numbers of people residing in one household and their relationships). She then described the

Careers in Sociology

Because sociologists attempt to understand, lessen, or solve problems that affect entire societies, they can choose among hundreds of career paths. Teaching and research—two activities chosen by Marta Tienda—remain the best-known callings of the sociologist.

Sociologists can also find jobs in business, the health professions, the criminal justice system, community organizations, and government. They may work closely with economists, anthropologists, psychologists, or political scientists.

An undergraduate degree—a B.A. (Bachelor of Arts)—in sociology will give you a strong base on which to build a career in business, social services, politics, or journalism. A sociology major is also good training for careers that require advanced degrees, such as law, education, medicine, or counseling.

An advanced degree—an M.A. (Master of Arts), or a Ph.D. (Doctor of Philosophy)— opens even more possibilities. A sociologist with one or both of those degrees might go on to become a college professor, a human resources manager, an urban planner, a criminologist, or a foundation president.

similarities and differences in family structure, education levels, poverty levels, and employment among the various groups of Hispanics living in the United States.

Marta's reports set the benchmark—the standard against which all other studies are judged—for Hispanic population studies. Marta's work put her on the map, too. She was fast becoming known as one of the country's top young sociologists.

~ The Voice of a People

Another opportunity to cement her reputation beckoned after the 1980 Census of Population and Housing—the once-every-10-years count of the U.S. population. To gauge the overall status of Spanish-speaking immigrants, Marta and one of her former University of Texas professors, Frank Bean, examined the Hispanic population of the United States. First, they scrutinized the findings of research others had done. Then they studied new evidence from the 1980 census, as well as data from earlier censuses. They looked at education, employment patterns, and economic characteristics such as annual income and poverty rates. These three areas tell sociologists whether a population is doing well.

By melding all this information, Frank and Marta were able to show changes that had occurred in the Hispanic population since 1960. And what they found was disturbing: The census comparisons told them that Hispanic Americans and Anglo-Americans were far from being equal in the schooling they received, in

¡USTED CUENTA!

en el censo de 1980

¡CUÉNTESE!

el 1 de abril

In 1980, United States census takers went door-to-door surveying household populations. Survey questions were printed in Spanish as well as in English.

U.S. CENSUS 1980

the kinds of jobs they held, and in the money they earned.

Frank and Marta coauthored a book based on their research. *The Hispanic Population of the United States* was the first authoritative description of the country's Hispanic population. Even though the book was published in 1987, it remains one of the best sources of information on Hispanics in the United States.

By speaking up for people whose voices had been too long ignored, Marta felt she was finally making a difference.

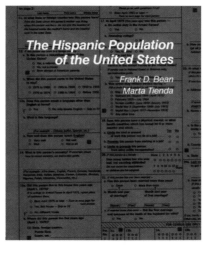

Frank Bean and Marta Tienda's study, published in 1987, is still regarded today as an excellent source of very detailed information about the U.S. Hispanic population.

Luis would nearly always

accompany

Marta on these

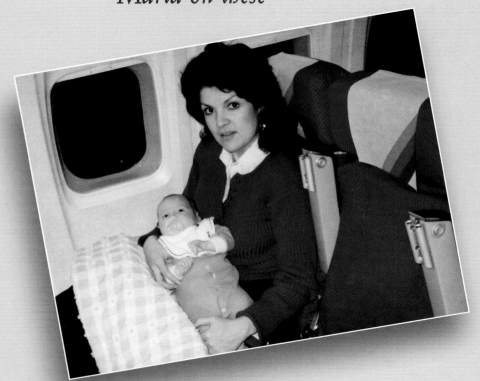

trips and soon
became a familiar sight
to Marta's colleagues.

ASKING WHY

10

I n 1982, Marta achieved a more personal landmark: Her son, Luis Gabriel, was born. Though Marta had child care for Luis while she taught her classes, she was called on frequently to travel. Luis would nearly always accompany Marta on these trips and soon became a familiar sight to his mother's colleagues. Luis was still only in diapers, yet he traveled all over the country and met college deans and administrators and foundation officers.

By the mid-1980s, Marta was growing restless. She had achieved a great deal in 11 years of teaching and research at the University of Wisconsin, but she was eager to accomplish much more. So many new challenges waited to be tackled! Marta began to seek them out.

Impressed by Marta's work, William Julius Wilson—head of the sociology department at the University of Chicago—offered her a teaching job in 1987. "You will thrive here," he assured her. "You'll work with others who are at the cutting edge of research in sociology."

Wilson was looking for a professor who would be bold, independent, and creative—a professor like Marta Tienda. She would be one of the nation's first sociologists to examine every major aspect of the Hispanic experience. Marta would focus on these people's ethnic and cultural diversity, their experience with the job market, how much they earned, how many lived in poverty,

At 2 months old, Luis Gabriel *(opposite)* was already becoming a frequent flyer. When Luis was almost 5, Marta began teaching at the University of Chicago *(above)*.

69

how many children they had, and countless other facets of their day-to-day existence. It was an opportunity tailor-made for Marta.

Wence, who was an independent businessman, saw the move as an opportunity to expand his business. So Marta, Wence, and Luis (who would soon be five years old) packed up and moved to Chicago.

~ Smart Style

Marta caused quite a stir when she strode into the dean's office at the University of Chicago for her first staff meeting. Amid the conservative bow ties and tweed jackets of her colleagues, Marta stood out with her mass of raven hair, denim jumpsuit, gold lamé belt, and gold lamé sandals. She talked her usual "mile-a-minute" as her bracelets chinked noisily against each other, causing a few professors to roll their eyes.

Marta was not about to back down just because her style might make some people uncomfortable. She has no use for stereotypes. In her world, the old notions of the nerdy natural scientist in his white laboratory coat or the dowdy social scientist in her frumpy clothes no longer apply. Marta may not be trying to make a statement, but she makes one anyway: It's okay to be smart and stylish.

Marta's mass of raven hair and stylish way of dressing caused quite a stir at staff meetings at the University of Chicago.

~ Three Ways to Slice the Data Pie

At the University of Wisconsin, Marta had been studying racial and ethnic groups in cross section. That means she had been using data from surveys (such as the census) to create "snapshots" of a certain population at one or several points in time. This kind of quantitative analysis, as it is called, shows how a population changes over time. Marta had found, for instance, that although different Hispanic groups often started from the same economic place, some succeeded more than others. Cubans gained the most economically, followed by Mexicans, but Puerto Ricans remained worse off. Fewer of them got jobs—and they didn't hold them as long as the other groups. Because of this, the income of Puerto Ricans living in the United States actually declined during the 1970s and 1980s.

Having answered what happened to these groups, Marta set out to discover why: Why did one group do better than another? At the University of Chicago, she began exploring methods that would let her follow the same people over time. These methods— together known as longitudinal analysis—would help her learn what had caused the changes she found in the cross section.

Finally, Marta carried out qualitative analysis, which hinged on interviewing people who had participated in the Urban Poverty and Family Life Survey. This kind of analysis puts a human face on statistics.

All three types of analysis—quantitative, longitudinal, and qualitative—helped Marta learn why certain patterns had developed in the data.

One pattern was the way three groups—Hispanics, blacks, and whites—made the transition from school to work. Marta found that each group behaved in a way that influenced its long-term economic status.

> At the University of Chicago, she began exploring methods that would let her follow the same people over time.

Hispanics tended to drop out of school and get jobs right away. This boosted their earnings at first, but their lack of schooling cost them in the long run.

Blacks tended to stay in school longer than Hispanics. Once they started to look for work, however, they ran into racial discrimination that prolonged their job search. For both reasons, blacks did not earn as much money as whites with the same amount of education.

To find out why some Hispanic groups did better than others, Marta and two of her graduate students interviewed people in various Chicago communities.

Whites stayed in school longer, then took less time to find work. This was the best outcome because with more education and no discrimination they earned more money over time.

~ Generation Maps

Marta wondered what role families—especially family outlook—played in the educational success of immigrants. With the help of a Chinese American graduate student named Grace Kao, she studied first-, second-, and third-generation families of Hispanics, Asians, and blacks.

In first-generation families, both parents and children are born in another country. In second-generation families, the parents are born in another country, but the children are born in the United States. And in third-generation families, both parents and children are born in the United States.

Second-generation students, Marta and Grace found, generally did best in school. In explaining why, Marta easily could have been describing her own family: Like her father, foreign-born parents have high hopes for their children's education. That's why many come to the United States in the first place. Immigrant parents often find themselves at the bottom of the ladder in economic and social standing, but they have a strong belief that with hard work they and their children can climb that ladder.

The second-generation children have better English-language skills than children born abroad (the first generation). This helps them do better in school, just as Marta did.

By the third generation, however, immigrants have often assimilated, or adopted, mainstream American values. The most successful may no longer identify themselves as Mexican, Korean, or whatever their country of origin is. The less successful may have lost their drive to excel. Why does this happen? It's a question sociologists are still trying to answer.

~ The Makings of a Mentor

Besides research, one of the most important jobs of the university sociologist is to train the next generation of scientists. Marta took her mentor's role seriously; she demanded a great deal of the students she took under her wing.

Marta has a reputation among her students as someone who reads every word of every paper they write. In the mold of her own former mentor, Harley Browning, she doesn't sugarcoat her criticism. She has no patience for a student unwilling to revise his or her work until it can withstand any scrutiny.

Some students say this makes Marta intimidating. Yet it also explains her strong support: Once she can make out the seeds of a student's idea, she pushes that student to expand on the idea and make it bloom.

In 1996, one of Marta's graduate students, Emilio Parrado, accompanied her on a research trip to Colombia, where he expected to get a chance to see the country.

Wrong!

Marta *(below left)* celebrates with her student Rebecca Raijman, who just finished writing her Ph.D. dissertation (a long research paper). With them are fellow graduate students Jeff Morenoff *(center)* and Emilio Parrado.

He and Marta worked from early every morning until late every night. On the return flight, Emilio saw her reading the airline magazine and assumed she was finally relaxing.

Wrong again!

Marta was feverishly highlighting passages in an article and jotting notes in the margins.

Like Emilio, Marta's other students had come to expect that when they gave her something to read, it would come back covered in red ink. "I'm known for spilling a lot of ink," Marta admits. "One time after I had graded a senior thesis, the student gave me a gift of three red pens tied up in a ribbon!" Yet there was a consolation: Marta's students knew that she marked up the papers of her colleagues just as much.

~ *Three Martas*

A proud Luis hugs his newborn brother, Carlos, in September 1989.

In 1989, Marta and Wence had another son, Carlos. With two children and a demanding career, Marta sometimes felt she had to be in two or more places at the same time. From 1994 until 1996, when she ran the University of Chicago's sociology department, she did seem to be everywhere at once: editing the *American Journal of Sociology*, where social scientists publish their research; serving on committees and boards to share her expertise on Hispanics, poverty, and employment; and testifying before government agencies and judges.

Once Marta testified in a housing-discrimination case in which a builder had tried to evict some Mexican residents from their apartments in order to demolish the buildings for a new development. Marta's testimony helped the residents win their case. Those who had been evicted and their buildings torn down got compensated for their loss. The others were allowed to stay in their homes, and their buildings were saved from demolition.

How could one person do so many things—and be in so many places—at the same time?

This question about the multiple movements of Marta Tienda inspired a skit about her at the University of Chicago's yearly Faculty Follies. In this show, graduate students got a chance to poke fun at their professors. Had you been in the audience, this is what you would have seen:

There was Marta—or someone who looked a lot like her—talking to a student, probably helping him see ways to improve a paper he had written.

But wait—there was Marta too, leading a faculty meeting! At least this second person looked like Marta, with big hoop earrings, high heels, and a short skirt. And, like Marta, she was talking extremely fast.

When a third Marta appeared on stage, dressed exactly like the other two, the audience burst out laughing. Marta—the real Marta Tienda—laughed along with them.

Of course it's impossible, but those who know Marta Tienda will insist she truly can be in three places at once!

Now it was
Marta's turn to
experience

what immigrants
go through.

11

THE IMMIGRANT EXPERIENCE

I magine leaving your native country and going to live in another one, with a new set of laws, a different language with its own alphabet, and customs that seem strange to you. Now imagine going to school in that country. It seems nearly impossible, but immigrants do it every day. Now it was Marta's turn to experience what immigrants go through.

Marta wasn't actually immigrating to another country. But she and her sons—4-year-old Carlos and 11-year-old Luis—were going to live in Israel for three months in 1994 while Marta wrote a book with her former graduate student, Israeli sociologist Haya Stier. Haya was working at Tel Aviv University, so Marta used that as an opportunity to get to know the country.

Marta and her husband, Wence, had been arguing constantly about priorities and finances, so he stayed in the United States. Sadly, their marriage was just about over. Always something of a dreamer, Wence had pursued several business schemes that lost a lot of money. After working so hard to climb out of poverty, Marta could no longer live with the fear of losing everything. She hoped this trip to Israel would prove that she could make it on her own.

An Orthodox Jew prays at the Western Wall (also called the Wailing Wall) in Jerusalem *(opposite)*. Customs such as this were new to Marta when she visited Israel. Above, Marta and Carlos stand at the border checkpoint between Jerusalem and Israel.

~ Stranger in a Strange Land

The Tiendas' first introduction to life in Israel was one of hunger. They had arrived in Tel Aviv, Israel's second largest city, on a Saturday. Many things in Israel shut down for the Sabbath, beginning at sundown on Friday and continuing until sundown on Saturday. The hungry family could get nothing from their hotel's room service that required cooking, so they settled for two pieces of bread and two glasses of juice. Luckily, Marta had brought a jar of peanut butter from home.

Small aggravations frustrated Marta daily. She sometimes had a hard time finding what she wanted in supermarkets. "Grocery day

Luis and Carlos pause near the entrance to the ruins of an amphitheater in Caesarea, an ancient city in Palestine.

this week was not pleasant," she wrote in the journal she kept while in Israel. "I am becoming more frustrated at the grocery store because there is little selection, I cannot read Hebrew, and thus can't sort sour cream from cottage cheese from plain yogurt. I have experienced a sense of humility for lack of skills in Hebrew. It is the utmost dependency to have to rely on others for communication." Marta could now appreciate what her father must have felt when he started working as a young man in Texas, unable to speak or understand English.

~ School Days

While Marta was working at Tel Aviv University, the boys attended Israeli schools. Carlos went to nursery school. Luis, who attended fifth grade back home in Chicago, was placed in sixth grade simply because the fifth grade class was full.

As a single working mother, Marta had a dilemma: What should she do about Carlos, who got out of school at 1:30 P.M.?

Her solution was to recognize that Luis was mature enough to escort his little brother home and care for him until 2:45, when Marta returned. To sweeten the deal, she paid Luis $2.00 for every hour he looked after Carlos.

Virtually all of the scientists and others with whom Marta dealt on a professional level spoke English, so no language barrier hindered her work. It was another matter for the boys at school, where the classes were conducted in Hebrew.

Carlos found himself in a large and boisterous class. Nearly 35 four- and five-year-olds competed for attention by yelling, overseen by a teacher given to yelling herself.

Luis had a different challenge: keeping up with more advanced work in a language he couldn't hope to learn in just a few months. He did his best and was able to make some headway in mathematics, where the language of numbers and symbols is universal.

~ The Poor Tell Their Story

During the day, while the boys were in school, Marta analyzed data and drafted chapters of the book she and Haya were coauthoring. *The Color of Opportunity: Pathways to Family, Welfare, and Work* is about the people—mainly minorities—who live in Chicago's inner-city ghettos. Marta had not only helped design the survey by coming up with the questions, but she also became the co-investigator, which means she helped lead the study. The data had been collected from a survey of nearly 2,500 people. The survey was considered trailblazing because it required the researchers to venture into dangerous, high-poverty neighborhoods, where they interviewed the residents face-to-face. *(See box, page 81.)* They asked such questions as:

"Do you consider yourself of Mexican or Puerto Rican descent? If yes, which one?"

"What race do you consider yourself to be, white or black?"

Published in 2001, *The Color of Opportunity* is the culmination of Marta and Haya's work studying minorities living in Chicago's inner city.

Haya Stier and
Marta Tienda

The Color of Opportunity

PATHWAYS TO
FAMILY, WELFARE,
AND WORK

"Did you graduate from high school?"

"As far as you know, what is your mother's highest level of education?"

"As far as you know, during the time you were growing up until you were about 14 years old, did your family ever receive public aid?"

Marta and Haya analyzed the survey data in search of answers to these questions: Do the poor who live in ghettos differ from other poor city residents in how they start families? In whether or not they receive welfare? In how they find and keep jobs? Marta and Haya found that the accumulated disadvantages of a lifetime cause poor urban minorities to differ significantly from people who are not poor. But they discovered only minor variations between the urban poor, who lived in very poor neighborhoods, and the poor in general. People who live in poor neighborhoods don't do well because they have fewer opportunities in education and the job market. Writing up their findings, Marta and Haya were able to document the many ways in which the color of one's skin continues to limit economic opportunity in the United States.

Haya and Marta's study on urban poverty in Chicago required researchers to venture into inner-city public housing projects to interview the residents.

~ Captivating the Crowd

Marta often spent her days working on the book and her nights writing journal articles. She also lectured about her research at Tel Aviv University. One talk in particular—about how Puerto Ricans fare in the U.S. job market—captivated the crowd. Despite all the

Choosing a Sample

When sociologists want to find information about a certain population of people, they almost always base their findings on a sample —that is, a portion of the population.

Why? Because the time and cost involved in gathering information about an entire population would make a survey nearly impossible to complete.

So how do sociologists choose their sample? Sometimes it is picked randomly, in which case it is assumed to represent on a small scale all the characteristics of the larger group. Samples can also be selected in a way that allows researchers to study the impact of certain characteristics—race or education, for example—on a population.

The Urban Poverty and Family Life Survey that was the basis for Marta and Haya's book, *The Color of Opportunity,* combined these two approaches. To begin, the survey wanted to include only people who lived in tracts (neighborhoods) with poverty rates of 20 percent or more. Researchers then further subdivided the people in that sample into four racial/ethnic groups: blacks, whites, Mexicans, and Puerto Ricans. A random sample of tracts was then chosen from within each racial/ethnic group.

This gave the researchers 330 tracts for the study. They then narrowed that sample to 125 tracts. From those 125 tracts, the researchers randomly selected and screened for eligibility (looking at age, sex, ethnicity, and parent status) to about 21,000 dwellings. The final sample included nearly 2,500 people, all of whom the researchers interviewed in person.

People living in low-income neighborhoods in Chicago *(above)* were the population from which Marta and Haya Stier drew their samples.

numbers and statistics Marta cited, many in her audience were genuinely moved as she described the obstacles that Puerto Ricans and other minorities routinely face in finding and holding jobs.

~ *History Comes Alive*

At the Wailing Wall in the Jewish quarter of Jerusalem, people come to pray. Others write their prayers on pieces of paper that are then stuck in the cracks.

Marta and her sons traveled throughout the Holy Land during their three months in Israel. They visited Masada, site of a mass suicide of cornered Jewish rebels who had fought the Romans about 2,000 years ago. They visited the Dead Sea, a lake six times saltier than the ocean and the lowest point on Earth.

During their last month, they made it to Jerusalem, Israel's capital and holy city to three religions—Judaism, Islam, and Christianity. The wind blew a cold rain—ideal weather for the homesick Chicagoans. Marta thought Jerusalem's Old City, with its narrow, cobbled streets crowded with people, was like taking a trip back in time.

Their host, Israeli social scientist Judah Matras, took them to the Jewish quarter of the city to see the Wailing Wall, the remnant of a great temple destroyed two millennia ago. While the boys went with their host to a prayer room off-limits to women, Marta marveled at the ways people can find to earn money. There was a service at the wall that, for a fee, took prayers faxed in from around the world and stuck them in the cracks and crevices between the wall's stones.

~ Embracing Differences

The trip to Israel gave Marta a vivid experience of life as a single parent. Though the task of balancing work and parenting some-times seemed overwhelming—especially when her sons got sick or were injured—she knew that it was something she could handle.

Now, after nearly three months, it was time for the family to pack up and return to their home in Chicago—but not before they had attended parties thrown for Luis, for Carlos, for Marta, and for all three. Marta held Israel and its people dear. As her journal made clear, she

Marta twirls Carlos at one of the farewell parties thrown for them before they left for the U.S.

had learned to embrace one more set of cross-cultural differences:

"Now I understand why the country closes on Saturday—it is the day for making and deepening friendships, and must not compete with other functions of life. Although this flustered me at first, I now believe it's a good custom to protect. Either I am turning into an Israeli or I have learned that life is meant to be enjoyed as well as lived."

"In the end," she says,
"we have one
primary

responsibility: our family."

MARTA'S DREAM

12

Inside a stone building on a leafy New Jersey campus, six young people sit around a polished table, listening intently to Marta Tienda. They are sociology students at Princeton University, and Marta is their teacher. She has been here as a professor and researcher since 1997, when she moved from Chicago to improve her family's quality of life. Worried about the impact of city living on Luis and Carlos, Marta had begun searching for a place with good public schools and safe streets. Princeton, with its leafy avenues, stately homes, and top-notch schools, more than fit the bill.

Indeed, of all Marta's accomplishments, being a single parent is to her the defining one. The challenge of being both mother and father to her two boys makes her work even more demanding. She calls it a "crazy conflict," but family always comes first.

When her ex-husband Wence died of cancer in 2001, Marta canceled all travel for three months so she would be available to her sons. When Luis suffered from depression during his freshman year in college, Marta took herself off committees to be more available to him. Nor is she afraid to call on family and friends when she needs help caring for her sons. "In the end," she says, "we have one primary responsibility: our family."

Luis, Marta, and Carlos enjoy playing with macaws during a trip to Cancun, Mexico *(opposite)*. Above, Marta rejoices when Carlos scores a goal in a soccer match at Princeton.

Moving east was a savvy career move for Marta. Princeton University is considered one of the most respected and prestigious colleges in the world. Sometimes Marta leads a seminar for a small group of graduate students there, teaching them how to conduct quantitative research. Like her mentor Harley Browning 30 years before, she guides them in fashioning testable questions and writing up the results.

One beautiful fall day, Marta led her six students in a discussion of the research proposals they had written for one of her courses. Each proposal must clearly state a research question—what does the student want to discover? She prodded each of her students with more questions:

"Why is this question important?"

"What have others done in search of the answer?"

"What will you do to find the answer?"

When one of the students seemed thrown by Marta's tough cross-examination, she made it clear that learning to give and receive criticism is an important goal of the course. It was partly Dr. Browning's tough critiques of her work, Marta knows, that made her strive to become the respected sociologist she is today. She wants nothing less for her own students.

Marta challenges her students in the finer points of conducting research.

One student read a proposal filled with jargon, or trendy academic terms. "You need to be able to communicate in simple English," Marta told him. "Sometimes people write in confusing ways to hide a lack of substance."

Wow! thought Marta. *That sounded just like Harley!* She clearly recalled his advice about her own less-than-stellar writing when she was just starting out: "If you want to be a sociologist," he had admonished her, "you'll need to learn how to write clearly and forcefully. A single well-written sentence is worth a thousand pictures."

Some of Marta's seminar students seemed overwhelmed by the difficulty of their tasks. Again harking back to when she was starting out in sociology, she reassured her students, "This will all seem a lot easier when you have 30 years of experience."

~ *Leveling the Playing Field*

Since 1999, Marta's research has focused on how colleges admit students. Many universities have been abandoning their policy of countering discrimination by giving special weight to an applicant's race and ethnicity, a practice called affirmative action.

As Marta sees it, affirmative action is about opportunity—finding qualified minority students and giving them a chance at a college education they might not otherwise get. She rejects the argument that it lowers standards by letting in unqualified students.

Marta is researching how the Texas Top 10 Percent Law is affecting student populations of such schools as the University of Texas at Austin.

As Marta points out, she herself is a product of affirmative action. Without the scholarships for Hispanics and other minorities that helped her attend graduate school, she would not be teaching the next generation of social scientists at Princeton today. But a lawsuit brought by a rejected white applicant in the mid-1990s caused the University of Texas (where Marta earned her doctorate) and other public universities to scrub their affirmative-action policies in 1996.

In response, the state of Texas crafted a plan to maintain student diversity at its public universities. The Texas Top 10 Percent Law guarantees admission to a state university to any

SOPHOMORE QUESTIONNAIRE
A. COURSE TAKING/GRADES

1. On a <u>typical day</u> during your sophomore year, how much time on average do you spend on homework in school and out of school? *(Fill in one response for each line)*

	Zero hours	Less than 1 hour	1 to 2 hours	3 to 4 hours	5 or more hours
a. Homework in school	○	○	○	○	○
b. Homework outside of school	○	○	○	○	○

2. When you are in the classroom do you... *(Fill in one response for each line)*

	Very rarely	Sometimes	Often	Almost all the time
a. Concentrate on what you are doing	○	○	○	○
b. Think about other things	○	○	○	○
c. Do more than is required	○	○	○	○
d. Stay focused on your class work	○	○	○	○
e. Daydream a lot	○	○	○	○

3. When you graduate, which of the following graduation plans will you be completing? *(Fill in one response only)*

○ Regular (general curriculum)
○ Recommended (college prep)
○ Distinguished Achievement
○ Don't know

4. Which of the following courses does your high school currently offer? *(Fill in one response for each line)*

5. Counting the courses you are taking this term, which of these courses have you taken, are you currently taking, or did not take? *(Fill in one response for each line)*

	Have already taken	Currently taking	Did not take
a. Algebra 1	○	○	○
b. Algebra 2	○	○	○
c. Geometry	○	○	○
d. Pre-Calculus	○	○	○
e. Calculus	○	○	○
f. Biology	○	○	○
g. Chemistry	○	○	○
h. Physics	○	○	○
i. Foreign Language (Spanish, French, other)	○	○	○
j. Computer Science	○	○	○

6. At the most recent grading period, what was your <u>grade</u> in each of the following subjects? *(Fill in one response for each line)*

	A	B	C	Lower than C	Didn't take the subject	Wasn't graded this way	Don't know
a. English or Language Arts	○	○	○	○	○	○	○
b. Mathematics	○	○	○	○	○	○	○
c. History or Social Sciences	○	○	○	○	○	○	○
d. Science	○	○	○	○	○	○	○

7. Which of the following Pre-Advance Placement (PAP) or Advance Placement (AP) courses does your high school currently offer? *(Fill in one response for each line)*

A sample page from the questionnaire created by Marta and her research team. Data collected from this survey will help them study the Texas Top 10 Percent plan.

Texas high school graduate who finishes in the top 10 percent of his or her class.

Marta and Teresa A. Sullivan, one of her former dissertation advisers from the University of Texas, are studying whether the plan is boosting or busting minority college enrollment in the state. Marta and her research team devised a questionnaire and gave it to a sample of almost 34,000 Texas high school sophomores and seniors. They will then follow the students' progress through high school and college.

~ Mixed Results

The study results paint a mixed picture for the Texas Top 10 Percent plan. In the plan's favor, Marta notes, it has given students all over the state a shot at attending the best public colleges that

Texas has to offer. Before the plan was enacted, some poor and rural high schools rarely sent a student to the University of Texas at Austin or Texas A&M—the state's top two public flagship schools.

But the plan is not perfect. It tends to overemphasize one factor—class rank. As a result, says Marta, the University of Texas at Austin is "drowning in top-10 percenters." The plan gives the universities less flexibility in shaping their freshman classes as they see fit.

Another argument against the plan is that it floods flagship schools with minorities who are ill-prepared for college, excluding better-qualified students who happen to rank just below that talented tenth. These people worry that the law will cause some of Texas's smartest students to leave the state and attend school elsewhere.

In the classroom, Marta passes on the passion she feels for her work. Her personal experiences and professional training put her in an excellent position to investigate the issues surrounding the Texas Top Ten Percent Law.

But Marta has marshaled research to disprove these allegations. Reaching into her sociologist's toolkit, she formed testable questions to see whether the attacks on the plan rely on solid fact or simple opinion. She asked, "Is it true that those who were ranked in the second 10 percent of their class—that is, those who attend competitive high schools but were denied automatic admission—are more likely to go to college out of state?"

To answer this question, Marta and her colleagues then examined the survey data that, at the second interview, recorded which students enrolled in college and where. The answer, they found, was no. Some students do leave the state, but the evidence that Marta gathered demonstrated they do so by choice, not because they are forced out. In other words, they would have gone out of state in any case.

In testimony before the Texas state legislature in June 2004, Marta used her research data to argue that the endangered Top Ten Percent plan should remain in force. But she did more than shoot down criticisms of the plan as misguided. She also recommended a way to make the law fairer. Admitting 50 rather than 70 percent of their entering freshmen using only the top 10 percent measure, Marta suggested, would leave the schools room to admit a more diverse student body. Schools could admit students who excelled not just in academics but in music and sports and leadership as well.

Not everyone welcomes Marta's outspokenness: She has received hate mail from parents whose children were rejected by their first-choice school. That's not going to stop Marta, though, in her quest to give all children an equal opportunity to get into college.

~ Spreading the Word

For Marta, being a sociologist is much more than doing research and reporting on it. She also speaks about her work to universities and organizations all over the country. By getting the word out about the importance of equal opportunity and access to good schools, she hopes that others will use her research to change things for the better.

In 2003, Marta's reputation as one of the top sociologists in the United States inspired the National Academies to ask her to lead a study on Hispanics in the United States. Hispanics are now the largest minority in this country, and their numbers are growing.

The committee's report describes how Hispanic families and communities are doing in several

Marta, shown here with then First Lady Hillary Clinton in 1997, is recognized as one of the top sociologists in the United States. She realizes the importance of speaking out for equal opportunities, especially when it involves education.

areas, like jobs, health care, and school. It will focus in particular on the needs of Hispanic children and young people, since helping them now would improve their lives as adults.

One of the disturbing things that the committee will talk about in the report is that Hispanic children who are born in the United States and grow up here are not as healthy as Hispanic children who are born in another country and then come here to live. The difference is their diets. Hispanic children who are born in the United States tend to eat a lot of fast food. They eat less fresh fruit and vegetables than their relatives in Latin America. As a result, a lot of them become overweight, even at a young age. This puts them at risk of developing other health problems, like diabetes and heart disease.

In contrast, Hispanic children who are immigrants tend to eat less fast food and more healthy foods like tortillas and beans as well as pineapple, bananas, and other fruits and vegetables. Thus, even though the United States is a rich country, living here has its risks. This is similar to what Marta Tienda and Grace Kao found in their study of immigrants and education.

Marta believes that addressing the needs of Hispanic children now will improve their lives as adults.

~ Knowledge Is Empowering

Marta is involved in a number of research projects that address the importance of education among minorities. Her findings reflect her own personal history: "Quality education is the most important factor in whether kids make it."

During a recent trip to her old neighborhood, Marta and her sister, Irene, stopped by their childhood home on Hartwick Street *(below)*. Marta also visited Lucille Page, the woman who always believed in her *(bottom)*.

That goes double, Marta knows, for poor kids who come from homes without books and magazines—homes like the one Marta grew up in. Like Marta, many Hispanic students have parents who did not finish high school; this puts the students at a competitive disadvantage in preparing for preschool reading and math. It also decreases their chances of learning about college, much less enrolling.

Marta was lucky. Her father, Toby Tienda, believed that a good education was the key to a better life. Today, Marta is pursuing that dream, which benefits children who are as disadvantaged as she once was.

Toby continues to chase his dream, too. After moving back to Texas in 1981, Marta's father finally returned to school, striving to earn his high school degree at the age of 74. During a business trip to

At age 74, Marta's father, Toby *(above)*, continued to chase his own dream of earning a high school diploma.

Texas in the summer of 2004, Marta spent an afternoon revisiting her childhood haunts with her father. As Toby steered his new GMC pickup truck through the streets of South Texas, Marta looked out at the people going about their everyday business on

92

the streets and in the stores. As she watched, she reflected on her life's journey. Then she turned and looked proudly at the man sitting next to her—the man whose firm belief in the power of a good education had inspired Marta to work so hard in school.

The girl who knew only that she wanted to help people had found a way to make a difference in society. She was indeed doing what Lucille Page had urged her to do so many years ago: Marta Tienda was passing it on.

TIMELINE OF MARTA TIENDA'S LIFE

1950   Marta Tienda is born on August 10.

1951   The Tienda family moves from Texas to Detroit, Michigan.

1957   Marta's mother, Azucena, dies of complications from surgery.

1959   Marta's father, Toribio, marries Beatrice Smoot.

1968   Marta graduates third in her class from Lincoln Park High School. She enters Michigan State University in East Lansing on scholarship.

1971   During the summer, Marta works for Michigan's State Department of Agriculture, certifying migrant workers for food stamps. She helps establish a day-care center for migrant children.

1972   Marta earns a bachelor's degree in Spanish from Michigan State University. Awarded a Ford Foundation Fellowship, Marta begins graduate studies at the University of Texas at Austin.

1974   Marta travels to Mexico for a summer seminar on women's roles in Latin America.

1975   Marta earns a master's degree in sociology from the University of Texas.

1976   Marta earns a Ph.D. in sociology from the University of Texas at Austin. Marta's brother, Juan Luis, is killed in an auto accident the day before her wedding to Wence Lanz in Texas. Marta and Wence move to Madison, Wisconsin, where she begins teaching sociology at the University of Wisconsin.

1982   Marta's son, Luis Gabriel, is born on November 22.

1987   *The Hispanic Population of the United States*, a research study coauthored by Frank Bean and Marta Tienda, is published. Marta and her family move to Chicago, where Marta begins her job as professor of sociology at the University of Chicago.

1989   Marta's son, Carlos, is born on September 18.

1994   Marta heads the Department of Sociology at the University of Chicago.
        She travels to Israel to work with colleague Haya Stier at the Tel Aviv
        University. Marta and Haya begin writing *The Color of Opportunity:*
        Pathways to Family, Welfare, and Work, which is published in 2001.

1997   Marta and her sons move to New Jersey, where Marta is Professor of
        Sociology and Public Policy at Princeton University.

1998   Marta directs Princeton's Office of Population Research until 2002.

2002   Marta serves as president of the Population Association of America, a
        nonprofit organization that promotes research on population issues.

2003   Marta begins a two-year term on the board of the Federal Reserve
        Bank of New York, part of the Federal Reserve System. The National
        Academy of Sciences asks Marta to lead a two-year study on
        Hispanics in the U.S. In October she is named one of the Top 100
        Influential Hispanics by *Hispanic Business Magazine.*

2004   Marta testifies before the Texas state legislature supporting the state's
        Top Ten Percent Plan for college admissions and recommending ways
        to make the plan fairer. In November she receives the Lifetime
        Achievement Award from Hispanic Business Inc.

2005   Marta continues to teach and serves on the boards of several
        philanthropic organizations.

About the Author

Award-winning journalist Diane O'Connell is the author of five books, including another biography in this series, *Strong Force*. She first became interested in sociology while working on the book *Divorced Dads: Shattering the Myths*, which she coauthored with Sanford L. Braver, Ph.D. Based on the largest federally funded study of divorced families in the United States, the book—and Diane—were featured on the news program 20/20. Before writing books, she was on staff at Sesame Workshop as a writer and editor for *Sesame Street Magazine*. Diane lives in New York City with her husband Larry and their golden retriever, Palmer.

GLOSSARY

This book is about a sociologist. To understand scientific words, it helps to know a little Greek and Latin. The word *socio* comes from the Latin *socius* meaning "companion" and *–logia* meaning the "science of." A *sociologist* is a scientist who studies the collective behavior of organized groups of people.

Here are some other words you will come across as you read about Marta Tienda's work as a sociologist. For more information about each word, consult your dictionary.

affirmative action: a practice that gives special consideration to minorities and women when it concerns school admission and employment. The action is designed to compensate for past discrimination against these groups.

assimilate: to become like the people of a particular nation; to adopt the cultural traditions and ways of a group.

average: a level (of intelligence, employment, education) typical of a group of people; the midpoint between extremes.

census: an official count or survey of a given population. The U.S. government conducts a census every ten years.

Chicano: an American of Mexican descent

Hispanic: a person living in the United States who is from a Spanish-speaking country

condition: to shape people's behaviors by repeatedly exposing them to a specific set of conditions or situations

cross section: a subset group of people who share the same characteristics as the larger group to which they belong; a representative sampling of a group of people.

data: facts and numbers, like statistics, gathered from scientific studies

demography: the science dealing with the statistics associated with human population. This can include birth rates, death rates, population size, education, and employment. From the Greek *demos*, meaning "people."

discrimination: a difference in the attitude toward and the treatment (usually unfair) of a particular group of people because of their race, religion, nationality, or gender. From the Latin *discriminatum* meaning "separated."

ethnic: belonging to a racial or cultural group. From the Greek *ethnos* meaning "nation."

ghetto: part of a city where members of a minority group live as a result of social or economic conditions

hypothesis: an idea that can be tested by observation or experiment

longitudinal analysis: a study of a group or groups of people over a long period of time

qualitative analysis: using data to identify reasons why populations change or act in a specific way. Qualitative data deals with meanings of things. It can be expressed through actions, words, or pictures, such as through interviews about one's job or family life.

quantitative analysis: using data from surveys, such as the census, to show how populations have changed. Quantitative data deals with numbers, such as population counts or employment rates.

sample: a part or portion of something that shows what the rest of it is like, such as a sample of high school students

social science: a branch of science that studies people and how they relate to each other as members of society, including their activities and institutions

stereotype: an oversimplified mental image or opinion about members of a group and their characteristics

statistics: a branch of mathematics that collects, classifies, and analyzes numerical data

urbanization: the development of cities and towns. From the Latin *urbanus* meaning "city."

Metric Conversion Chart

When you know:	Multiply by:	To convert to:
Miles	1.61	Kilometers
Acres	0.40	Hectares
Gallons	3.79	Liters
Kilometers	0.62	Miles
Hectares	2.47	Acres
Liters	0.26	Gallons

FURTHER RESOURCES

Women's Adventures in Science on the Web

Now that you've met Marta Tienda and learned all about her work, are you wondering what it would be like to be a sociologist? How about a planetary astronomer, a forensic anthropologist, or a robot designer? It's easy to find out. Just visit the *Women's Adventures in Science* Web site at **www.iWASwondering.org**. There you can live your own exciting science adventure. Play games, enjoy comics, and practice being a scientist. While you're having fun, you'll also get to meet amazing women scientists who are changing our world.

BOOKS

Atkin, S. Beth. *Voices from the Fields: Children of Migrant Farmworkers Tell Their Stories.* Boston: Joy Street Books, 1993. The personal stories of nine Mexican American children and dramatic photographs depict the harsh realities of the lives of migrant farmworkers and their families.

Bandon, Alexandra. *Mexican Americans.* New York: Maxwell Macmillan International, 1993. First-person narratives focus on the recent history of Mexican immigration to the United States.

Kotlowitz, Alex. *There Are No Children Here: The Story of Two Boys Growing Up in the Other America.* New York: Anchor Books, 1992. A powerful account of two brothers, Lafayette and Pharoah Rivers, who grew up in the Henry Horner housing project in Chicago. This story personalizes the data that Marta Tienda collected on urban poverty and family life.

Morey, Janet and Wendy Dunn. *Famous Mexican Americans.* New York: Cobblehill Books, 1989. This book showcases the achievements and contributions of 14 Mexican Americans from various professions.

WEB SITES

Carnegie Corporation of New York:
http://www.carnegie.org/reporter/08/interview/interview_low.html
Marta Tienda served for eight years as a trustee of the Carnegie Corporation of New York, a nonprofit foundation. In this interview for the *Carnegie Reporter*, she discusses her educational opportunities, how she chose a career as a demographer, and why she focuses her research on minorities and higher education.

Scholarships for Hispanics: http://www.scholarshipsforHispanics.org/
The National Education Association and National Hispanic Press Foundation sponsor this Web site. It makes more than 1,000 sources of financial aid accessible to Hispanic students and offers guidelines on how to search and apply for scholarships.

Texas Top 10% Project: http://www.texastop10.princeton.edu/
This site features an interview with Marta Tienda about her current research projects, including the Texas Top 10% project to find out how changes in admission policies affect minority college enrollment in Texas.

United States Census Bureau: http://www.census.gov/
Look up current statistics and other information about the social and economic characteristics of the Hispanic (Latino) population.

SELECTED BIBLIOGRAPHY

In addition to interviews with Marta Tienda, her family, friends, and colleagues, the author did extensive reading and research to write this book. Here are some sources she consulted.

Bean, Frank D. and Marta Tienda. *The Hispanic Population of the United States.* New York: Russell Sage Foundation, 1987.

Kanellos, Nicolás, ed. *The Hispanic-American Almanac: A Reference Work on Hispanics in the United States.* Detroit: Gale Research, 1993.

Stier, Haya and Marta Tienda. *The Color of Opportunity: Pathways to Family, Welfare, and Work.* Chicago: University of Chicago Press, 2001.

Tienda, Marta and Julius Wilson, eds. *Youth in Cities: A Cross-National Perspective.* Cambridge: Cambridge University Press, 2002.

Index

A
Accents, understanding, 15
Advanced degrees, 65
African American studies, 71
Alpena County, Michigan, 1, 37, 39
American dream
    believing in, 12
    expanding, 3–4
Anglo-Americans, 3, 71–72
Auto industry
    employment in, 9, 11–12, 15, 19
    strikes in, 19

B
Babies, cost of, 59–61
Bachelor of Arts (B.A.) degree, 65
Bean, Frank, 66–67
Benchmarks, Marta establishing, 66
Browning, Harley, 55–57
    as *maestro*, 48–52
    teaching Marta the value of writing
        clearly, 55, 73, 86

C
Cancun, Mexico, 85
Careers, in sociology, 65
Census of Population and Housing of
        1980, 66
Charity, accepting, 17
Chicago, 72–73, 79–81, 83
Child-protection agency, 16
Clear writing, learning the value of, 55, 86
Clinton, Hillary, 90
Collaborations, between Juan Luis and
        Marta, 23, 33, 64–65
College
    degrees awarded in, 65
    Marta's teacher opening her eyes to,
        27–29
    second-generation students in, 72
    See also individual colleges and
        universities
The Color of Opportunity, 79, 81
Cost of babies, to society, 59–61
Creativity, Marta valuing, 35

Crop picking, 4, 8, 19–20
    See also Migrant workers
Cuernavaca, Mexico, 56

D
Data analysis, 25, 50–51, 71
    getting answers from, 53
Day-care, for migrant workers, 41–43
Demography, 53–54, 59
Deportation, Mexicans facing risk of, 8
Doctor of Philosophy (Ph.D.) degree, 65

E
Education
    Marta spreading the message about,
        85–93
    providing greater opportunities in, 4

F
Faculty Follies, skit about Marta, 75
Farm laborers. *See* Migrant workers
Fast food, 91
Feminist movement, 56
Food stamps, 1–2, 37–39
Ford Foundation scholarship, 45–46, 55
Foster care, fighting off, 16
Free worker housing, 39–40

G
Generations, mapping, 72–73
Girls' Athletic Association (GAA), 31–32
Goodfellows organization, 17

H
Harshness, children uniting against,
        22–23
Hepatitis, 25
The Hispanic Population of the United States, 67
Hispanic population studies, 67, 69,
        90–91
    and children's health, 91–92
    establishing benchmark for, 66
    of Michigan, 47
    transition from school to work for, 71
    See also Latin American men; Mexican
        Americans
Hualahuises, Mexico, 7
Huff Junior High School, 27
Human populations. *See* Demography

I
Immigrants
  getting help from family, 9
  as parents, 72–73
  sharing feelings of, 77–83
Independence, Marta seeking, 31–34
Israel, Marta's trip to, 76–83

J
Jews, Orthodox, 77
Jobs
  father losing, 19, 61
  held by Marta, 37–43, 47, 62, 69, 85
  See also Careers

K
Kao, Grace, 72, 91

L
Labor force, Mexican women in, 54
Language barriers, 40, 78–79
  See also Accents
Lanz, Carlos (son), 74, 77–79, 83, 85
Lanz, Luis Gabriel (son), 68–69, 74,
  77–79, 85
Lanz, Wence (husband), 59, 61–62, 64,
  70, 77, 85
Latin American studies, 48
Lincoln Park, Michigan, 15–16, 34, 48
Lincoln Park High School, 31, 33–34
Literature, connection with sociology, 49
Longitudinal analysis, 71

M
Mapping generations, 72–73
Master of Arts (M.A.) degree, 65
  thesis for, 53–55
Math, unlocking the secrets held by
  numbers, 25
Matras, Judah, 82
Max Paun Elementary School, 13, 27
Mexican Americans, 3–4
  first-ever national survey of, 64
  identity issues for, 45–46, 56–57
  offering nutritional advice to, 47
  resentment toward, 39
  sources of information about, 67
Mexican Revolution, 19
Mexico, women in the labor force in, 54

Michigan Department of Agriculture, 37
Michigan Department of Social Services,
  42
Michigan State University, 1, 33–35,
  45–47, 63
Migrant workers, 8, 20, 37, 39–43
  day-care for, 41–43
  free housing for, 39, 40
  legal help for, 63
Monroe County, Michigan, 19–20
"Moonlighting," 11
Mortar Board Society, 45

N
National Honor Society, 32
Night school
  father going to, 12
  Lucille Page going to, 23
Numbers
  holding secrets, 25
  as used by sociologists, 49
Nutritional advice, 47, 91

O
Orthodox Jews, 77

P
Page, Lucille (neighbor)
  advising Marta, 48, 92–93
  loaning Marta money, 64
  never judging Marta, 23
Parrado, Emilio, 73–74
Paz, Octavio, 49
Pino, Frank, 45
Population Research Center, 52–53, 55
Population studies, establishing
  benchmark for Hispanic, 66
Poverty
  among urban minorities, 80
  in countries with high population
  growth rates, 59–61
Princeton University, 85–86, 89
Problem-solving ability, 43, 90
Puerto Rican Americans, in the U.S. job
  market, 80–82

Q

Qualitative analysis, 71
Quantitative analysis, 71
Questions, fashioning testable, 52, 86, 89

R

Raijman, Rebecca, 73
Random samples, 81
Research methods, 50, 52, 86
Resentment
    toward Marta, 90
    toward Mexican Americans, 39
Resources, sending where needed, 53
Rights, to life, liberty, and the pursuit of
    happiness, 3
Rio Grande (river), Mexicans crossing, 1, 7
Risk-taking, 11
Rossi, Rudee, 33–34, 37
Rules, going wrong, 39–41

S

Sabbath, The, as observed in Israel, 78
Sampling, 81
Scholarships, 28, 33, 45
Scientific method, 25
Second-generation students, in college, 72
Shoes, symbolism of, 17–19
Single parenting, 85
Smoot, Beatrice (stepmother), 21
    divorcing Toby, 46
    harshness toward the children, 21–23,
        31–33
    marrying Toby, 18–19
Social science, 2–3, 48
Sociology
    careers in, 65
    connection with literature, 49
    eighth grade research project in, 28
    numbers used in, 48–49
    of the soul, 1–5
Socorro, Grandma, 14–15
Standard deviation, 51
Statistics, 50–51
Steinem, Gloria, 49
Stier, Haya, 77, 79–81
Strikes, 19
    See also Union rules
Sullivan, Teresa A., 88
"Surplus commodities," 17

T

Teaching certification, 35
Tel Aviv University, 77–78, 80
Testable questions, fashioning, 52, 86, 89
Texas Top 10 Percent Law, 87–90
Tienda, Azucena (mother), 12
    birth of her first three children, 9
    death of, 14
    marrying Toby, 8–9
Tienda, Gloria and Irene (younger sisters),
        12, 18–19, 21–22, 92
Tienda, Juan Luis (brother), 12, 21–23,
        25–26
    birth of, 9
    death of, 62–64
    interest in Chicano studies, 45
    "partner in crime" with Marta, 23, 33
Tienda, Maggie (older sister), 7, 11–12,
        14, 18–19, 21–22, 32
Tienda, Marta
    ambition to excel, 34
    aspiring to be a beautician, 26–29
    assessing the Texas Top 10 Percent
        Law, 87–90
    birth of, 9
    births of her sons, 69, 74
    collaboration with Juan Luis, 23, 33,
        64–65
    death of her mother, 14–15
    disillusioned with teaching, 35, 43
    early poverty of, 50
    establishing benchmark for Hispanic
        population studies, 66
    experiences picking crops, 19–20
    Faculty Follies skit about, 75
    feeling different, 13–14
    first trip to Mexico, 16
    in graduate school, 55, 61
    graduating from Michigan State
        University, 46–47
    jobs held by, 37–43, 47, 62, 69, 85
    learning English, 11
    learning the value of writing clearly,
        55
    marriage coming to an end, 77
    marrying Wence Lanz, 58–59, 61–62
    mentoring her students, 73–74
    as a problem solver, 43, 90

publishing *The Color of Opportunity*, 79, 81

publishing *The Hispanic Population of the United States*, 67

punished in elementary school, 13–14

recalling her past, 4–5, 61, 78, 92–93

receiving athletic recognition, 31–32

remembering Juan Luis, 62–64

resentment toward, 90

responsibilities while growing up, 21–22

seeing herself as brown, 45–46

seeking independence, 31–34

setting up day-care for migrant workers, 41–43

sharing feelings of what immigrants go through, 77–83

speaking Spanish, 12, 34–35

stepmother entering her life, 19

teachers influencing, 27–29, 35, 45, 52

teaching and living in Israel, 77–83

valuing creativity, 35

wanting to help others, 35, 49

Tienda, Reynaldo (brother from stepmother Beatrice), 21

Tienda, Toribio "Toby" (father), 1, 14–15, 21

dreaming of a better life for his family, 5, 7–9

going to school, 12, 92–93

hard work of, 11–12

losing his job, 19, 61

marrying Azucena secretly, 8–9

second marriage, to Beatrice Smoot, 18–19, 46

seeking a better life, 9

setting the standard for excellence, 5

valuing education, 24, 92–93

Transition from school to work, 71–72

"Tweeners," 56

U

Undergraduate degree, 65

Union rules, 11

University of Chicago, 69–71, 74–75

University of Michigan, 63

University of Texas at Austin, 48, 52–53, 87, 89

University of Wisconsin, 62, 64, 69–70

Urban Poverty and Family Life Survey, 71, 81

U.S. Border Patrol, 7

U.S. Department of Labor, 65

U.S. job market, Puerto Rican Americans in, 80–82

"Us against them" attitude, 36, 38–39, 41

V

Vietnam War, 63

W

Wailing Wall (in Jerusalem), 76–77, 82

Whites. *See* Anglo-Americans

Wilson, William Julius, 69

World War II, 7

Writing clearly, learning the value of, 55, 86

WOMEN'S ADVENTURES IN SCIENCE ADVISORY BOARD

A group of outstanding individuals volunteered their time and expertise to help the Joseph Henry Press conceptualize and develop the *Women's Adventures in Science* series. We thank the following people for their generous contribution of time and talent:

Maxine Singer, Advisory Board Chair: Biochemist and former President of the Carnegie Institution of Washington

Sara Lee Schupf: Namesake for the Sara Lee Corporation and dedicated advocate of women in science

Bruce Alberts: Cell and molecular biologist, President of the National Academy of Sciences and Chair of the National Research Council, 1993–2005

May Berenbaum: Entomologist and Head of the Department of Entomology at the University of Illinois, Urbana-Champaign

Rita R. Colwell: Microbiologist, Distinguished University Professor at the University of Maryland and Johns Hopkins University, and former Director of the National Science Foundation

Krishna Foster: Assistant Professor in the Department of Chemistry and Biochemistry at California State University, Los Angeles

Alan J. Friedman: Physicist and Director of the New York Hall of Science

Toby Horn: Biologist and Co-director of the Carnegie Academy for Science Education at the Carnegie Institution of Washington

Shirley Jackson: Physicist and President of Rensselaer Polytechnic Institute

Jane Butler Kahle: Biologist and Condit Professor of Science Education at Miami University, Oxford, Ohio

Barb Langridge: Howard County Library Children's Specialist and WBAL-TV children's book critic

Jane Lubchenco: Professor of Marine Biology and Zoology at Oregon State University and President of the International Council for Science

Prema Mathai-Davis: Former CEO of the YWCA of the U.S.A.

Marcia McNutt: Geophysicist and CEO of Monterey Bay Aquarium Research Institute

Pat Scales: Director of Library Services at the South Carolina Governor's School for the Arts and Humanities

Susan Solomon: Atmospheric chemist and Senior Scientist at the Aeronomy Laboratory of the National Oceanic and Atmospheric Administration

Shirley Tilghman: Molecular biologist and President of Princeton University

Gerry Wheeler: Physicist and Executive Director of the National Science Teachers Association

Illustration Credits:

JHP Executive Editor: Stephen Mautner

Series Managing Editor: Terrell D. Smith

Designer: Francesca Moghari

Illustration research: Joan Mathys

Special contributors: Meredith DeSousa, Allan Fallow, Mary Kalamaras, April Luehmann, Faith Mitchell, Mary Beth Oelkers-Keegan, Anita Schwartz

Graphic design assistance: Michael Dudzik